Characterization of the Hydrologic Resources of San Miguel County, New Mexico, and Identification of Hydrologic Data Gaps, 2011

By Anne Marie Matherne and Anne M. Stewart

Prepared in cooperation with San Miguel County, New Mexico

Scientific Investigations Report 2012–5238

U.S. Department of the Interior
U.S. Geological Survey

U.S. Department of the Interior
KEN SALAZAR, Secretary

U.S. Geological Survey
Marcia K. McNutt, Director

U.S. Geological Survey, Reston, Virginia: 2012

This and other USGS information products are available at http://store.usgs.gov/

U.S. Geological Survey
Box 25286, Denver Federal Center
Denver, CO 80225

To learn about the USGS and its information products visit http://www.usgs.gov/
1-888-ASK-USGS

Suggested citation:
Matherne, A.M., and Stewart, A.M., 2012, Characterization of the hydrologic resources of San Miguel County, New Mexico, and identification of hydrologic data gaps, 2011: U.S. Geological Survey Scientific Investigations Report 2012–5238, 44 p.

Acknowledgments

This study was conducted with the cooperation of San Miguel County, New Mexico. We would also like to thank Lani Tsinnajinnie and Catherine Lucero for their assistance with parts of this report.

Contents

Figures

Tables

Conversion Factors

Inch/Pound to SI

Multiply	By	To obtain
Length		
inch (in.)	2.54	centimeter (cm)
inch (in.)	25.4	millimeter (mm)
foot (ft)	0.3048	meter (m)
mile (mi)	1.609	kilometer (km)
Area		
square mile (mi^2)	259.0	hectare (ha)
square mile (mi^2)	2.590	square kilometer (km^2)
Volume		
acre-foot (acre-ft)	1,233	cubic meter (m^3)
acre-foot (acre-ft)	0.001233	cubic hectometer (hm^3)
Flow rate		
acre-foot per year (acre-ft/yr)	1,233	cubic meter per year (m^3/yr)
acre-foot per year (acre-ft/yr)	0.001233	cubic hectometer per year (hm^3/yr)
cubic foot per second (ft^3/s)	0.02832	cubic meter per second (m^3/s)
Hydraulic conductivity*		
foot per day (ft/d)	0.3048	meter per day (m/d)
Transmissivity**		
foot squared per day (ft^2/d)	0.09290	meter squared per day (m^2/d)
Storage coefficient***		
dimensionless (ft^3/ft^2/ft) or (ft^3/ft^3)	0.0001	dimensionless (m^3/m^2/m) or (m^3/m^3)

Temperature in degrees Fahrenheit (°F) may be converted to degrees Celsius (°C) as follows:

$$°C = (°F - 32)/1.8$$

Vertical coordinate information is referenced to the North American Vertical Datum of 1988 (NAVD 88).

Horizontal coordinate information is referenced to the North American Datum of 1983 (NAD 83).

*Hydraulic conductivity (K): Defined as T divided by the saturated aquifer thickness, foot squared per day per foot. In this report, the mathematically reduced form, foot per day (ft/d), is used for convenience.

**Transmissivity: The standard unit for transmissivity is cubic foot per day times foot of aquifer thickness per square foot [(ft^3/d × ft) / ft^2]. In this report, the mathematically reduced form, foot squared per day (ft^2/d), is used for convenience.

***Storage coefficient: The volume of water an aquifer releases from or takes into storage per unit surface area of the aquifer per unit change in head.

Characterization of the Hydrologic Resources of San Miguel County, New Mexico, and Identification of Hydrologic Data Gaps, 2011

By Anne Marie Matherne and Anne M. Stewart

Abstract

The U.S. Geological Survey (USGS), in cooperation with San Miguel County, New Mexico, conducted a study to assess publicly available information regarding the hydrologic resources of San Miguel County and to identify data gaps in that information and hydrologic information that could aid in the management of available water resources. The USGS operates four continuous annual streamgages in San Miguel County. Monthly discharge at these streamgages is generally bimodally distributed, with most runoff corresponding to spring runoff and to summer monsoonal rains. Data compiled since 1951 on the geology and groundwater resources of San Miguel County are generally consistent with the original characterization of depth and availability of groundwater resources and of source aquifers. Subsequent exploratory drilling identified deep available groundwater in some locations. Most current (2011) development of groundwater resources is in western San Miguel County, particularly in the vicinity of El Creston hogback, the hogback ridge just west of Las Vegas, where USGS groundwater-monitoring wells indicate that groundwater levels are declining.

Regarding future studies to address identified data gaps, the ability to evaluate and quantify surface-water resources, both as runoff and as potential groundwater recharge, could be enhanced by expanding the network of streamgages and groundwater-monitoring wells throughout the county. A series of seepage surveys along the lengths of the rivers could help to determine locations of surface-water losses to and gains from the local groundwater system and could help to quantify the component of streamflow attributable to irrigation return flow; associated synoptic water-quality sampling could help to identify potential effects to water quality attributable to irrigation return flow. Effects of groundwater withdrawals on streamflow could be assessed by constructing monitoring wells along transects between production wells and stream reaches of interest to monitor decline or recovery of the water table, to quantify the timing and extent of water-table response, and to identify the spatial extent of capture zones. Assessment of groundwater potential could be aided by a county-wide distribution of water-level information and by a series of maps of groundwater potential, compiled for each individual aquifer, including saline aquifers, for which the potential for municipal use through desalination could be explored. A county-wide geographic information system hydrologic geodatabase could provide a comprehensive picture of water use in San Miguel County and could be used by San Miguel County as a decision-support tool for future management decisions.

Introduction

The surface-water and groundwater resources of San Miguel County, New Mexico, are increasingly relied upon to meet growing municipal, domestic, livestock, and agricultural needs. San Miguel County is an area with an expanding economy and population, and to meet future water demands, aquifers may be further developed. Periodic dry periods further focus attention on the quantity and sustainability of the surface-water and groundwater resources. Since a study by Griggs and Hendrickson (1951), however, only a few published studies have focused on the hydrogeology and associated surface-water/groundwater interactions within San Miguel County. As a result, there are limited publicly available fundamental groundwater data (such as long-term groundwater levels) and surface-water data upon which to interpret hydrologic processes. The U.S. Geological Survey (USGS), in cooperation with San Miguel County, conducted a study to assess publicly available information regarding the hydrologic resources of San Miguel County and to identify data gaps in that information and additional hydrologic information that could aid San Miguel County's management of available water resources.

Purpose and Scope

This report characterizes the current (2011) state of knowledge of the hydrologic resources of San Miguel County, N. Mex., by using publicly available reports and existing sources of data. Streamflow data are presented for six streamgages, including two streamgages in the adjacent county, with respect to annual and monthly flow. Flow characteristics are discussed on the basis of daily and annual maximum discharge. Surface-water quality, based on New Mexico Environment Department assessments, is summarized. A general hydrostratigraphic framework for San Miguel County is developed on the basis of a compilation of previous work in the area. Well logs from the New Mexico Office of the State Engineer and USGS databases and groundwater levels from the USGS monitoring network are used to determine the current (2011) understanding of the hydrogeologic framework and groundwater conditions. On the basis of this characterization, critical data gaps in understanding and assessing the hydrologic resources of San Miguel County are identified, and options for future study, assessment, and monitoring are presented.

Physical Setting of San Miguel County

San Miguel County covers an area of about 4,750 square miles (mi²) in northeastern New Mexico (fig. 1A). The longest east-west dimension of the county is about 117 miles (mi), and the longest north-south dimension is about 57 mi. The county seat, Las Vegas, lies about 40 mi due east of the New Mexico State capital, Santa Fe (fig. 1A).

San Miguel County comprises four physiographic areas: the Sangre de Cristo Mountains in the northwest, Glorieta Mesa in the southwest, the Las Vegas Plateau and outliers, which includes escarpment areas, in the north-central and eastern areas, and the plains and southern hogback (monocline) east of the Sangre de Cristo Mountains and Glorieta Mesa (Griggs and Hendrickson, 1951) (fig. 1B). Elevations range from a high of about 11,800 feet (ft) above the North American Vertical Datum of 1988 (NAVD 88) in the Sangre de Cristo Mountains to a low of about 3,900 ft where the Canadian River exits the county (fig. 1A). Glorieta Mesa ranges in elevation from about 7,000 to 8,000 ft, and the Las Vegas Plateau ranges in elevation from about 5,000 to 6,800 ft. The topography declines steeply along the winding Canadian Escarpment to the plains, which range in elevation from about 4,000 to 5,000 ft (Griggs and Hendrickson, 1951).

The types of principal aquifers in the study area vary by physiographic area. In the northwest mountainous area, fractured granitic bedrock and localized shallow alluvial deposits are the primary sources of groundwater. On Glorieta Mesa, groundwater occurs primarily within Permian sedimentary formations, including the Yeso Formation, the Glorieta Sandstone, and the Triassic Santa Rosa Sandstone. Well depths in this area range from less than 100 ft below land surface (bls) to more than 1,100 ft bls (Griggs and Hendrickson, 1951). The major aquifer within the Las Vegas Plateau is the Cretaceous Dakota Sandstone. Depth to water is generally less than 250 ft bls. In the plains area, groundwater is derived primarily from the Triassic Chinle Formation and Santa Rosa Sandstone, at depths of about 100–300 ft bls.

San Miguel County is drained primarily by the Pecos and Canadian Rivers (fig. 1A). The Pecos River and its major tributaries originate in the Sangre de Cristo Mountains, as do the headwaters of the Sapello River, which flows into the Mora River in Mora County to the north. The Mora River flows into the Canadian River south of the San Miguel–Mora County line (fig. 1A). The principal tributaries of the Pecos River are, in downstream order, the Rio Mora (distinct from the Mora River), Bull Creek, Tecolote Creek, and the Gallinas River. The Gallinas River and Tecolote Creek drain to the east and southeast, cutting through a series of hogback monoclines, which are steep structural ridges in transition areas between the western mountains and the eastern plains. The valleys of these streams are generally narrow and confined. The Canadian River, in the eastern plains region, enters San Miguel County at the northern county border and flows in a southerly direction through the Canadian River canyon to Conchas Lake and then continues generally easterly below Conchas Lake to the eastern county line.

Temperatures in San Miguel County range from winter low temperatures below 0 degrees Fahrenheit (°F) in the Sangre de Cristo Mountains (Natural Resources Conservation Service, 2011a) to summer high temperatures greater than 100 °F in the eastern plains (Western Regional Climate Center, 2011). Mean monthly temperatures range from about 21–51 °F in the western mountains (1990–2011 period of record) (Natural Resources Conservation Service, 2011a) to about 39–81 °F around Conchas Dam (30-year normal climate record, 1981–2010) (Western Regional Climate Center, 2011).

Precipitation decreases from the mountains in the northwestern part of the county to the south and to the eastern plains. Mean annual precipitation, as recorded at six climate stations with long continuous records (longer than 30 years) and at one high-elevation Snowpack Telemetry (SNOTEL) station (22 years of record), ranges from about 40 inches per year at the Wesner Springs SNOTEL station in the Sangre de Cristo Mountains to about 13 inches per year at the Villanueva precipitation station (fig. 2 and table 1) (Natural Resources Conservation Service, 2011a; Western Regional Climate Center, 2011). Monthly precipitation at low-level to mid-level elevations is greatest during the summer monsoons, with an average of about 37 percent of mean annual precipitation occurring during July and August (fig. 3A–E and G). Precipitation at the Wesner Springs SNOTEL station, at an elevation of 11,120 ft, is bimodally distributed, with 27 percent of the mean annual precipitation occurring during July and August and 18 percent of the mean annual precipitation occurring during March and April (fig. 3F). The Wesner Springs SNOTEL station measures snowpack and snow-water equivalent, which is the amount of water contained within the snowpack (Natural Resources Conservation Service, 2011b). Mean monthly snow water equivalent at Wesner Springs for March and April is 13 inches (Natural Resources Conservation Service, 2011a). Spring snowmelt runoff generally occurs from April through June, and the irrigation season is April 1 through October 31 in San Miguel County (Aguirre, 2008). Responding to precipitation and snowmelt runoff, more than 80 percent of the mean annual discharge at the six streamgages occurs during the 7-month period from April through October (fig. 4A–F).

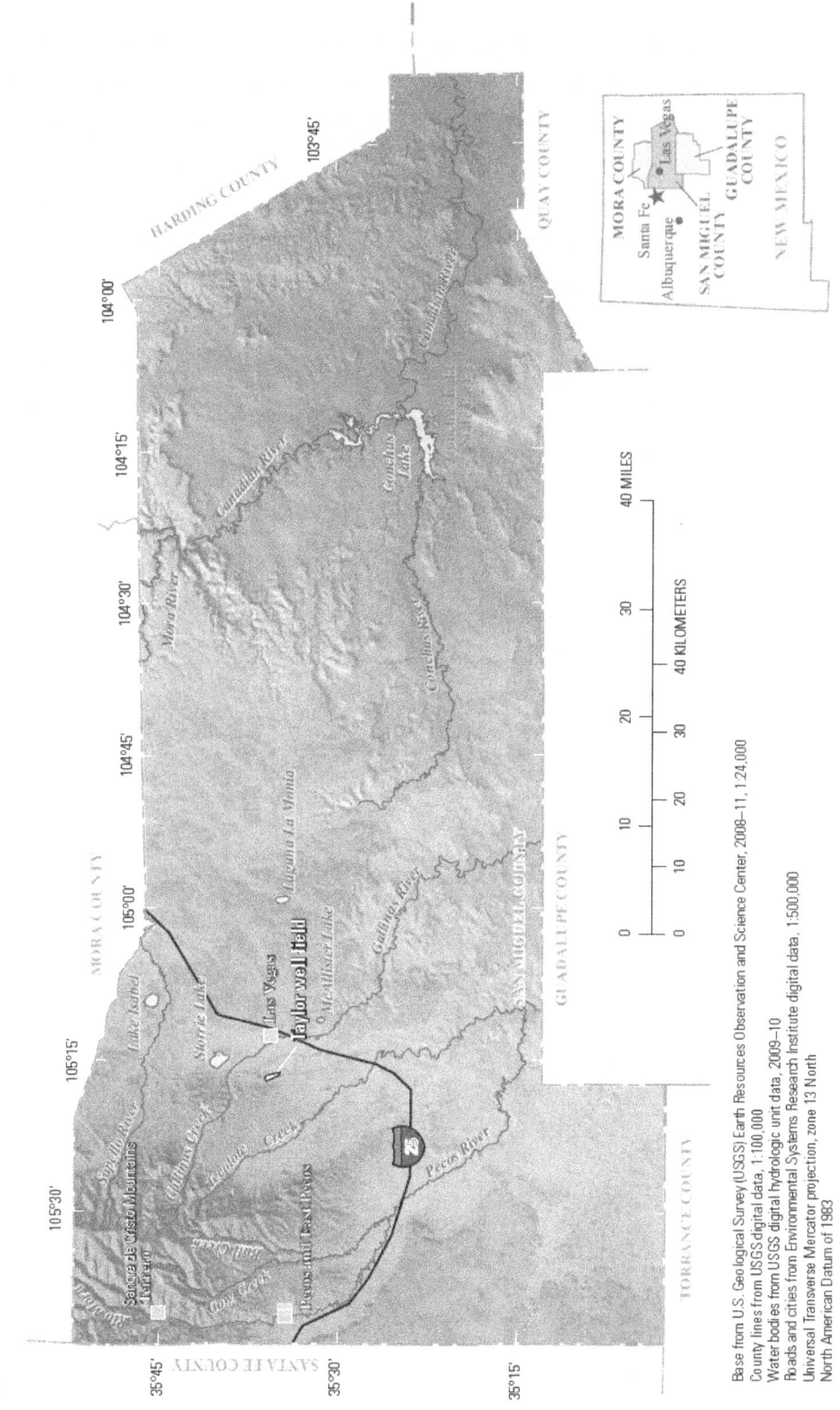

Figure 1. The study area of San Miguel County, New Mexico. *A,* Topography, major water bodies, roadways, cities, towns, and physical features. *B,* Physiographic areas as described by Griggs and Hendrickson (1951) and the Gallinas Creek area as described by Baltz (1972).

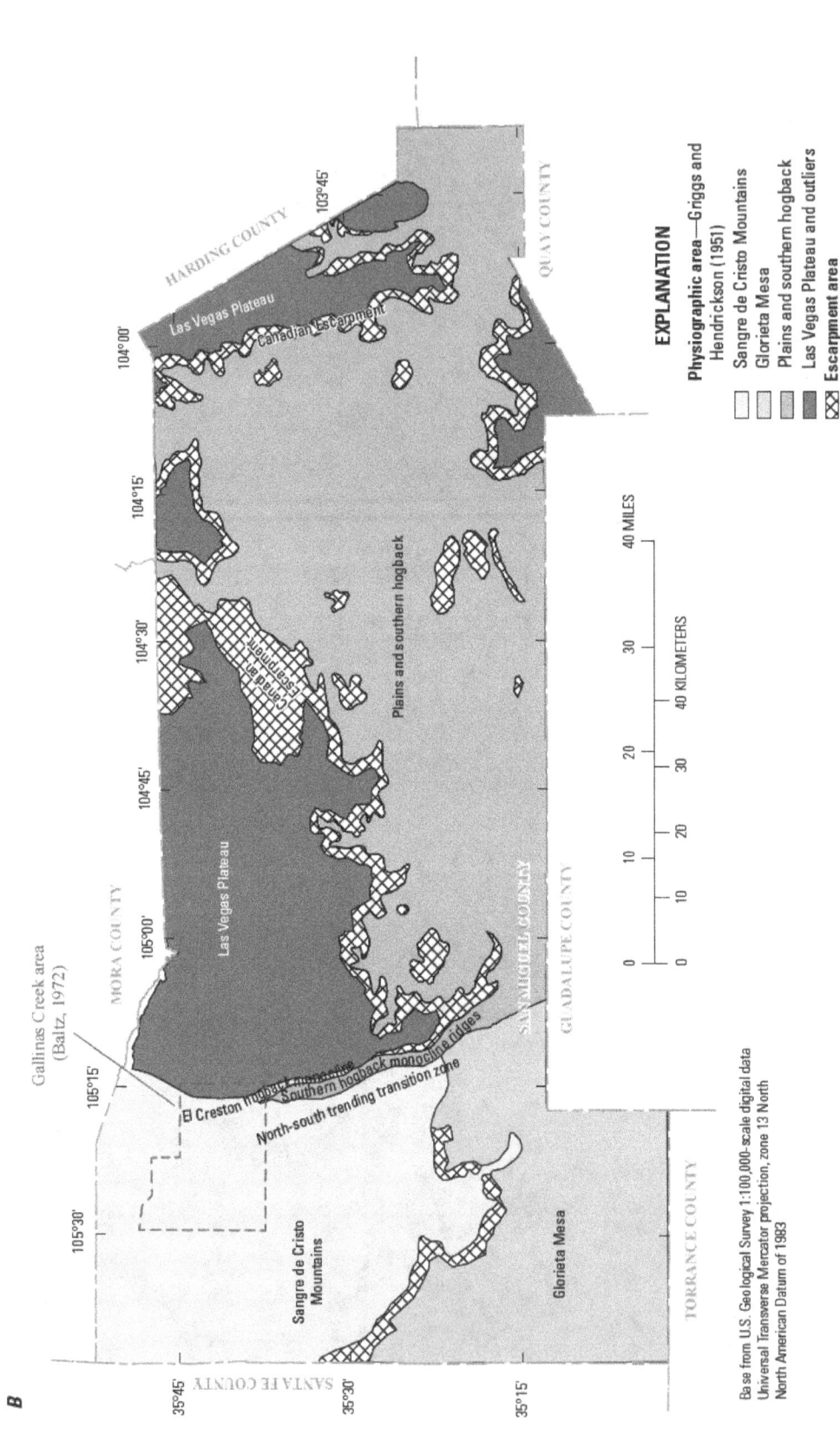

Figure 1.   The study area of San Miguel County, New Mexico. *A*, Topography, major water bodies, roadways, cities, towns, and physical features. *B*, Physiographic areas as described by Griggs and Hendrickson (1951) and the Gallinas Creek area as described by Baltz (1972).—Continued

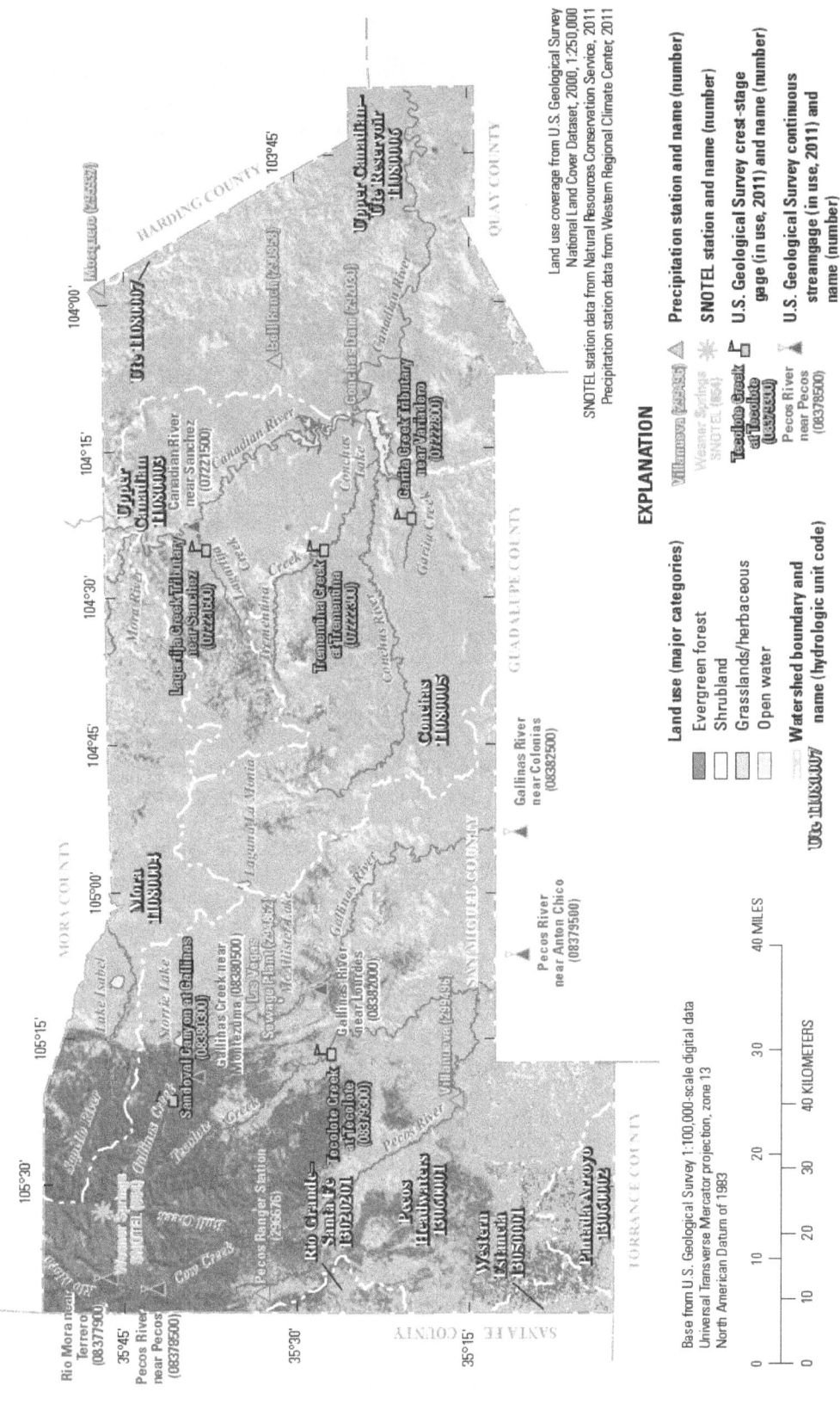

Figure 2. Major surface-water features and measurement sites, including precipitation and Snowpack Telemetry (SNOTEL) stations, crest-stage gages, and streamgages, in San Miguel County, New Mexico.

Table 1.   Location of selected National Weather Service precipitation stations and Natural Resources Conservation Service Snowpack Telemetry (SNOTEL) station in San Miguel County, New Mexico, and mean annual precipitation for the period of record.

[, degrees; ', minutes]

Site identification number (fig. 2)	Site name	Latitude	Longitude	Period of record	Mean annual precipitation (inches)
290858	Bell Ranch	35°32'	104°06'	1899–2010	14.89
292030	Conchas Dam	35°24'	104°11'	1936–2010	14.31
294862	Las Vegas Sewage Plant	35°32'	105°12'	1983–2010	18.05
296676	Pecos Ranger Station	35°35'	105°41'	1916–2010	16.30
295937	Mosquero	35°48'	103°56'	1915–2010	16.48
299496	Villanueva	35°16'	105°22'	1942–2010	12.67
854	Wesner Springs SNOTEL	35°47'	105°33'	1988–2010	39.95

Climate in New Mexico exhibits alternating wet and dry periods (Committee on the Scientific Basis of Colorado River Basin Water Management, 2007). In the latter decades of the 20th century, regional climate records show a tendency toward greater variability in precipitation and multiyear episodes of both wet and dry conditions. The early years of the decade beginning in the year 2000 were drought years (Committee on the Scientific Basis of Colorado River Basin Water Management, 2007). The 2011 water year (October 2010 through September 2011) was the driest year on record; statewide, precipitation was only 52 percent of normal (the mean precipitation value for the period 1980–2010) and, in San Miguel County, averaged 20–40 percent of normal (National Weather Service, 2011).

Seventy-four percent of the land of San Miguel County, about 3,500 mi², is agricultural, of which about 92 percent is in pasture (U.S. Department of Agriculture, 2007). Within the remaining, nonagricultural lands, there are two major population centers. The county seat, Las Vegas, had a 2010 population of about 13,750, or about 20,000 when the surrounding area was included (U.S. Census Bureau, 2010). The Village of Pecos had a 2010 population of 6,445 (U.S. Census Bureau, 2010). Projected population increase is primarily along the Interstate 25 corridor between Santa Fe and Las Vegas (fig. 1A) from individuals living in San Miguel County and commuting to Santa Fe for work and other activities (Daniel B. Stephens & Associates, Inc., 2005).

Surface water supplies about 97 percent of the water demand in San Miguel County. Of the total surface-water consumptive use in San Miguel County, 51 percent is used for irrigated agriculture, and 45 percent is lost to reservoir evaporation. About 3 percent serves public water supplies, and less than 1 percent serves commercial and livestock uses (Longworth and others, 2008). San Miguel County contains parts of five New Mexico Office of the State Engineer (NMOSE) declared underground water basins (fig. 5): the Upper Pecos, the Canadian River, and the Tucumcari underground water basins, with minor areas of the Northern Rio Grande and Estancia underground water basins included along the western county border. Seventy-two percent of groundwater withdrawals in San Miguel County supplies individual domestic wells and public water supplies, with the remainder divided between livestock and commercial uses (Longworth and others, 2008).

Both surface-water and groundwater rights in San Miguel County are managed by the NMOSE. Historically, acequias, or community ditch irrigation systems, are managed as principal local government units for the distribution and use of surface water (New Mexico Office of the State Engineer, 2011a). A total of 10,986 acres of land are irrigated with surface water in San Miguel County (Longworth and others, 2008). Surface water in the Pecos River and Canadian River is regulated under the Pecos River and Canadian River Interstate Stream Compacts (New Mexico Office of the State Engineer, 2011b). Conchas Lake, managed by the U.S. Army Corps of Engineers, provides storage for irrigation water near Tucumcari, N. Mex., about 35 mi to the southeast, and to the Bell Ranch, northeast of the reservoir (University of New Mexico, 2012).

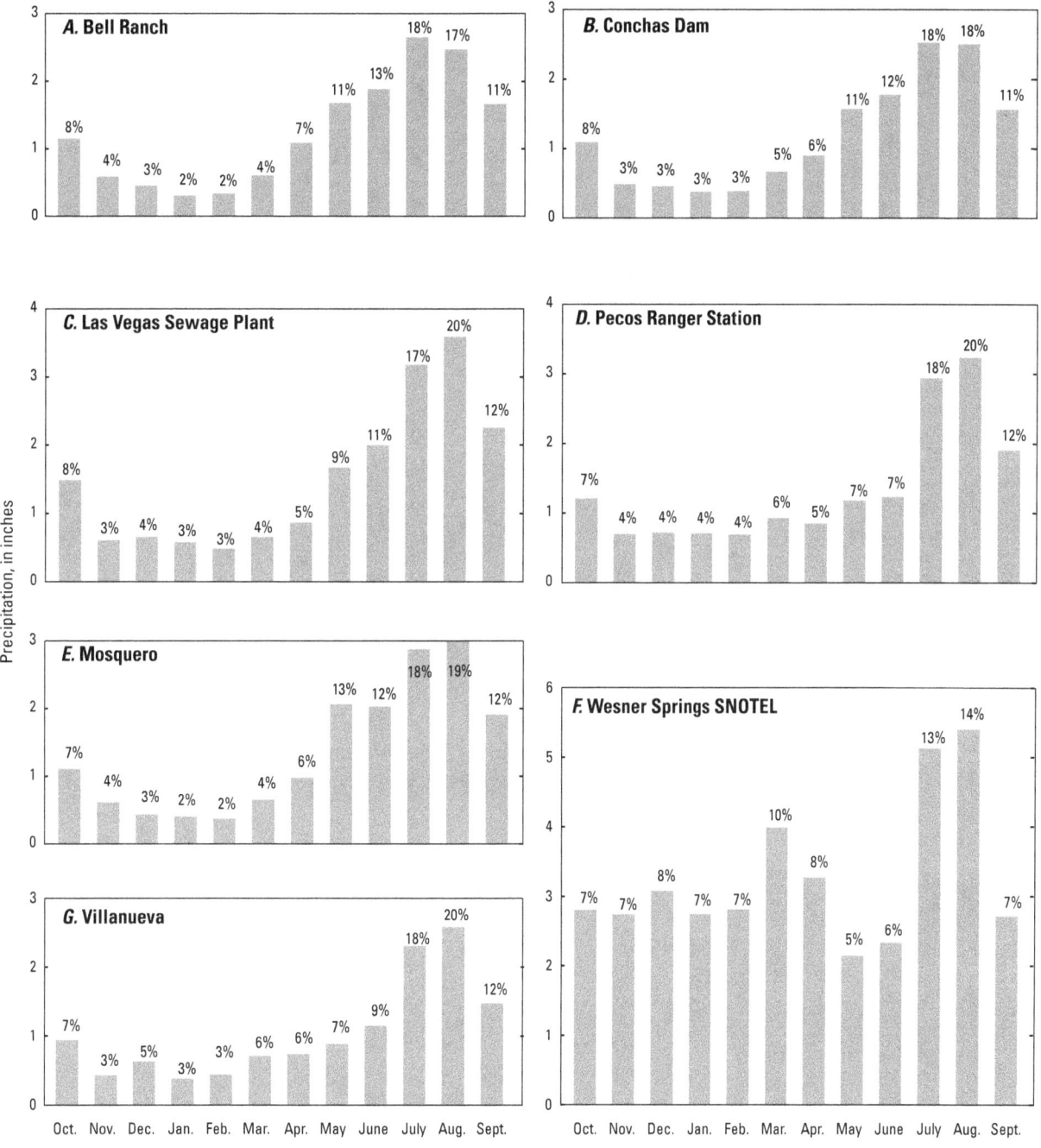

Figure 3.   Mean monthly precipitation for the period of record and percent of mean annual precipitation at climate stations in San Miguel County, New Mexico. *A*, Bell Ranch. *B*, Conchas Dam. *C*, Las Vegas Sewage Plant. *D*, Pecos Ranger Station. *E*, Mosquero. *F*, Wesner Springs Snowpack Telemetry (SNOTEL). *G*, Villanueva.

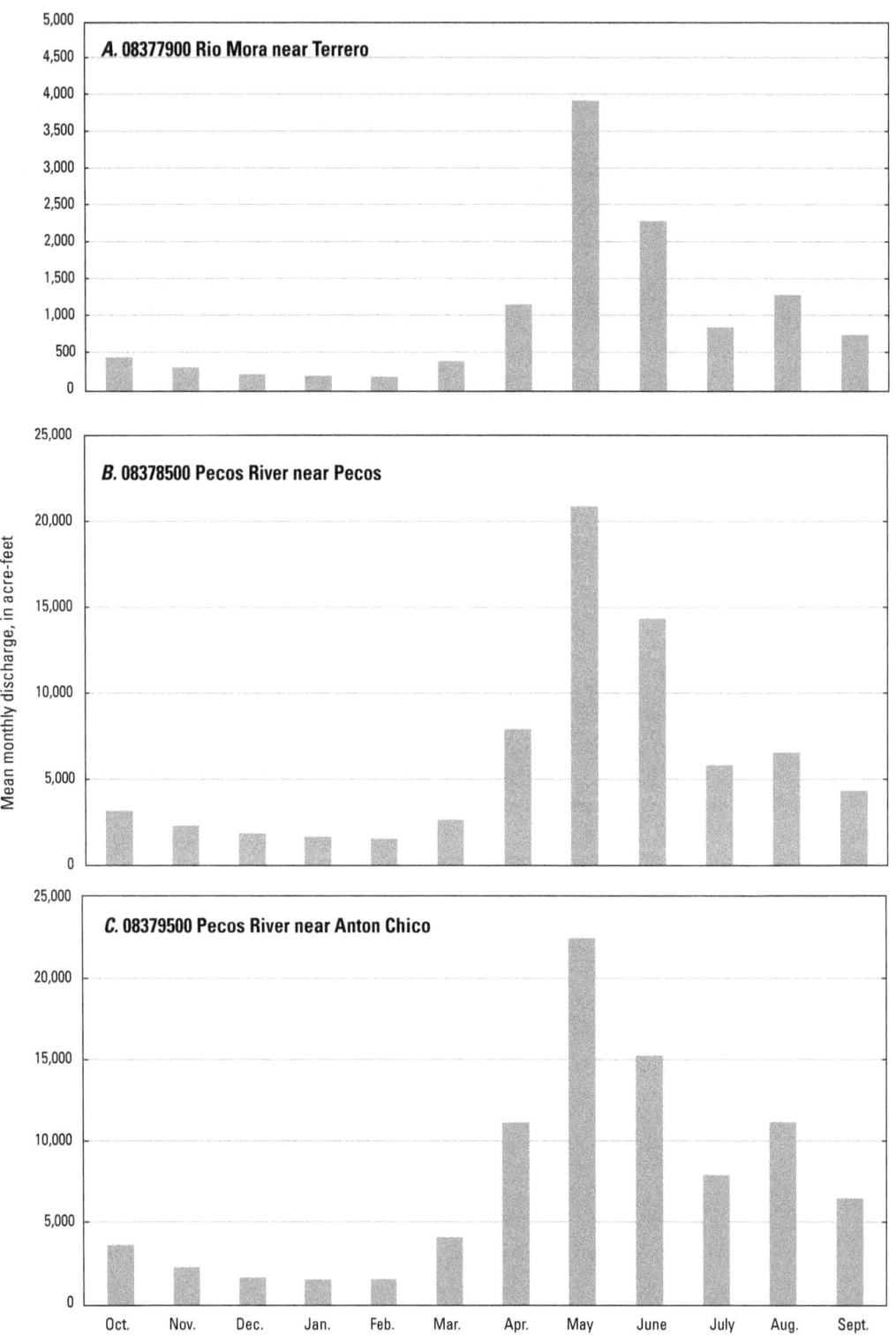

Figure 4. Mean monthly discharge for the period of record for six streamgages in San Miguel and Guadalupe Counties, New Mexico. A, Rio Mora near Terrero (08377900). B, Pecos River near Pecos (08378500). C, Pecos River near Anton Chico (08379500). D, Gallinas Creek near Montezuma (08380500). E, Gallinas River near Colonias (08382500). F, Canadian River near Sanchez (07221500).

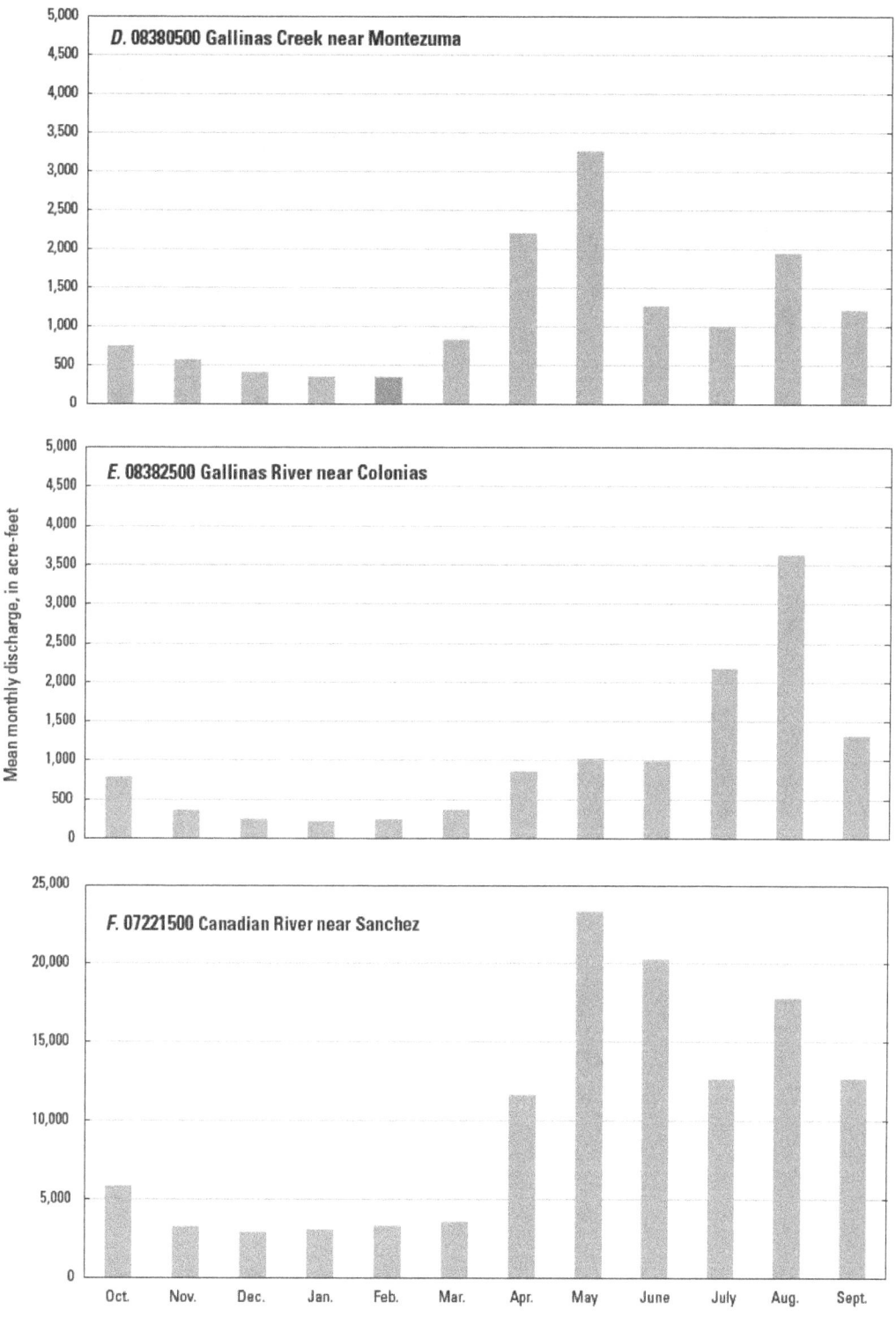

Figure 4. Mean monthly discharge for the period of record for six streamgages in San Miguel and Guadalupe Counties, New Mexico. *A*, Rio Mora near Terrero (08377900). *B*, Pecos River near Pecos (08378500). *C*, Pecos River near Anton Chico (08379500). *D*, Gallinas Creek near Montezuma (08380500). *E*, Gallinas River near Colonias (08382500). *F*, Canadian River near Sanchez (07221500).—Continued

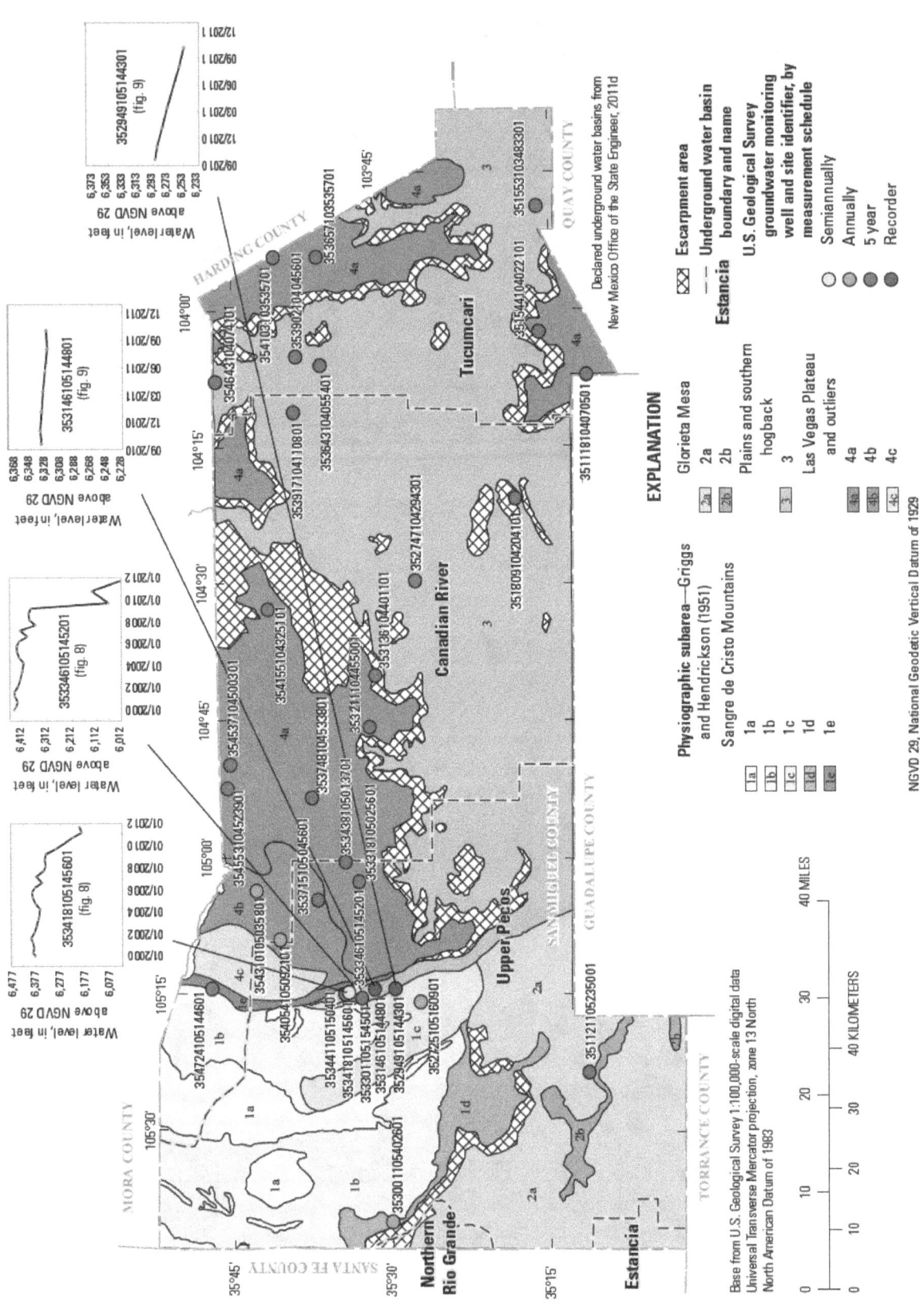

Figure 5. Griggs and Hendrickson (1951) physiographic areas and subareas, New Mexico Office of the State Engineer (2011d) underground water basins, locations of wells in the U.S. Geological Survey groundwater-monitoring well network, and hydrographs for selected U.S. Geological Survey groundwater-monitoring wells with periods of record longer than 2 years in San Miguel County, New Mexico.

Approach

Written Resources

Current understanding of the hydrologic resources of San Miguel County was summarized on the basis of a review of available literature and of existing groundwater and surface-water data. Written information used in characterizing the hydrologic resources was obtained from publicly available sources, including site-specific, privately prepared reports on file at the library of the New Mexico Office of the State Engineer. Written resources reviewed for this report are compiled in the bibliography.

Surface-Water Resources

Streamflow and peak discharge data for rivers and streams in San Miguel County were obtained from USGS streamgage and crest-stage gage records, archived in the USGS National Water Information System (NWIS) database (http://waterdata.usgs.gov/nm/nwis/sw). A total of 19 streamgages and 9 crest-stage (peak streamflow) gages have been operated by the USGS in San Miguel County (table 2). Five continuous streamgages are currently (2011) operated by the USGS on the major rivers in San Miguel County: Canadian River near Sanchez (07221500); Rio Mora near Terrero (08377900); Pecos River near Pecos (08378500); Gallinas Creek near Montezuma (08380500); and Gallinas River near Lourdes (08382000). The streamgage Gallinas River near Lourdes, in operation from 1952 to 1963, was reestablished in 2005 as a seasonal gage (April through September). Data from this station are not included in this discussion because current annual records are not available for comparison to the other station records. The stations Pecos River near Anton Chico (08379500) and Gallinas River near Colonias (08382500) (table 2), both in northern Guadalupe County, are included to represent downstream flow conditions on the Pecos and Gallinas Rivers. There are no current pairs of upstream and downstream streamgages for a similar comparison of the Canadian River. Peak streamflow is currently (2011) recorded at five crest-stage gages: Lagartija Creek Tributary near Sanchez (07221600); Trementina Creek at Trementina (07222300); Garita Creek Tributary near Variadero (07222800); Tecolote Creek at Tecolote (08379300); and Sandoval Canyon at Gallinas (08380300) (fig. 2 and table 2).

Data are presented in this report for the six continuous streamgages, including the two in northern Guadalupe County, in the form of monthly and annual hydrographs and flow duration curves. Flow duration curves characterize the distribution of annual flow at a streamgage in terms of the exceedance probability, the percentage of time a given discharge is equaled or exceeded. Annual maximum discharge for a range of recurrence intervals is also presented for the streamgages and crest-stage gages to characterize the occurrence of extreme flows at those locations.

The exceedance probability of discharge at a streamgage, on which the flow duration curves are based, was calculated for the 30-year period 1981–2010 on the basis of the mean daily discharge values by using the following equation (Dunne and Leopold, 1978):

$$P = 100 * [M/(n+1)] \tag{1}$$

where

$P =$ the probability, in percentage of time, that a given discharge will be equaled or exceeded;

$M =$ the rank value of the discharge, in descending order; and

$n =$ the number of observations for the period.

Estimates of instantaneous annual maximum peak discharges were calculated for a range of recurrence intervals at stream and crest-stage gages by using the USGS statistical program PeakFQ (Flynn and others, 2006), which fits a Pearson Type III frequency distribution to the logarithms of instantaneous annual peak flows for the period of record at a gage. The parameters of the Pearson Type III frequency distribution are estimated by the logarithmic sample moments (mean, standard deviation, and coefficient of skewness) with adjustments for low outliers, high outliers, historical peaks, and generalized skew. The recurrence interval of a peak discharge is based on the probability that an event of a given magnitude will be equaled or exceeded in any given year (U.S. Geological Survey, 2011a). A peak discharge with a 1 in 100 chance of occurrence in any given year is said to have a 100-year recurrence interval.

Groundwater Resources

Data on groundwater wells in San Miguel County, including well locations and water levels, were obtained from the New Mexico Water Rights Reporting System (NMWRRS) (http://nmwrrs.ose.state.nm.us/nmwrrs/index html) and the USGS NWIS (http://waterdata.usgs.gov/nwis/gw) databases. These data, and information regarding acquisition and reporting of the well records, are presented in Matherne and Stewart (2012). The NMWRRS dataset contained 2,176 well records for San Miguel County, of which only 11 did not include data for total well depth. The USGS NWIS database contained 317 well records for San Miguel County, of which 297 were cross-indexed with NMWRRS point-of-diversion (POD) identification numbers and for which tapped aquifers were identified. Because the NWIS dataset was, in general, a subset of the NMWRRS dataset, the larger NMWRRS dataset was used to locate well-completion depths, and the more detailed NWIS dataset was used to identify tapped aquifers.

Table 2. Location of streamgages and crest-stage gages operated by the U.S. Geological Survey in San Miguel and Guadalupe Counties, New Mexico, period of record, contributing drainage area of gaged stream, and 30-year mean annual discharge of selected streamgages.

[USGS, U.S. Geological Survey; mi², miles squared; , degrees; ', minutes; ", seconds; - -, not reported]

USGS site identification number	Site name	Period of record (water year)	Latitude	Longitude	Contributing drainage area of gaged stream (mi²)	30-year mean annual discharge 1981–2010 (acre-feet per year)
Streamgages						
07218700	Manuelitas Creek near Rociada	1957–63	35°49'30"	105°23'55"	52	
07220000	Sapello River at Sapello	1918; 1957–73	35°46'11"	105°15'05"	132	
07220100	Lake Isabel Canyon near Sapello	1965–74	35°44'42"	105°09'25"	- -	
07220600	Sapello River near Watrous	1957–63	35°46'05"	105°02'28"	213	
07221500	Canadian River near Sanchez	1913–14; 1937–2011	35°39'17.4"	104°22'43"	5,712	97,800
07222000	Canadian River near Bell Ranch	1930–38	35°30'00"	104°15'00"	5,900	
07222500	Conchas River at Variadero	1937–96	35°24'10"	104°26'35"	393	
07223000	Bell Ranch Canal below Conchas Dam	1971–84	35°24'10"	104°11'07"	- -	
07223300	Conchas Canal below Conchas Dam	1971–89	35°22'35"	104°10'03"	- -	
07224500	Canadian River below Conchas Dam	1937–38; 1943–72	35°24'32"	104°10'10"	6,984	
08377900	Rio Mora near Terrero	1964–2011	35°46'38"	105°39'29"	53.2	26,330
08378500	Pecos River near Pecos	1920; 1924; 1931–2011	35°42'30"	105°40'58"	189	79,100
08379178	Tecolote Creek at Wright Canyon near El Porvenir	1989–90 (18 months)	35°41'20"	105°28'49"	- -	
08379185	Wright Canyon at Mile 0.55 near El Porvenir	August 1990	35°41'44"	105°28'45"	- -	
08379187	Tecolote Creek below Wright Canyon near El Porvenir	summer months, 1989–92	35°40'19"	105°27'58"	5.42	
08379200	Tecolote Creek near San Pablo	1961–65	35°33'10"	105°22'10"	83	
08379500	Pecos River near Anton Chico (Guadalupe County)	1929–2011	35°10'43"	105°06'32"	1,050	90,730
08380500	Gallinas Creek near Montezuma	1927–2011	35°39'07"	105°19'08"	84	16,260
08381000	Gallinas Creek at Montezuma	1907–66	35°39'15"	105°16'30"	87	
08382000	Gallinas River near Lourdes	1952–63, 2005–11 (seasonal)	35°28'08"	105°09'41"	313	
08382500	Gallinas River near Colonias (Guadalupe County)	1952–2011	35°10'55"	104°54'01"	610	13,470
Crest-stage gages						**Number of peaks recorded**
07221600	Lagartija Creek Tributary near Sanchez	1972–2011	35°38'19"	104°24'56"	1.16	22
07222300	Trementina Creek at Trementina	1959–2011	35°28'02"	104°25'00"	64	49
07222800	Garita Creek Tributary near Variadero	1971–2011	35°20'09"	104°21'50"	12	36
08378000	Pecos River near Cowles	1911–19	35°45'08"	105°40'30"	189	9
08379300	Tecolote Creek at Tecolote	1937–2011	35°27'27"	105°16'41"	122	55
08379550	Canon Blanco near Leyba	1971–83	35°13'14"	105°40'12"	11.2	12
08379600	Pecos River Tributary near Dilia	1952–84	35°12'50"	105°04'50"	0.16	33
08380300	Sandoval Canyon at Gallinas	1957–2011	35°41'13"	105°21'30"	7.60	48
08381700	Canon Piedra Lumbre near Las Vegas	1971–75	35°34'14"	105°17'50"	8.06	5

Groundwater levels in San Miguel County are monitored as part of a groundwater-monitoring network maintained by USGS in cooperation with the NMOSE. Groundwater levels are generally measured by using a steel tape and by following standard USGS protocols (Cunningham and Schalk, 2011). Prior to 2010, the USGS groundwater-monitoring network in San Miguel County, maintained in cooperation with the NMOSE, consisted of seven wells: one well that was part of a statewide network monitored on a 5-year basis, two wells monitored semiannually, and four wells monitored annually. The network was focused in western San Miguel County, that part of the county with the greatest population density (U.S. Census Bureau, 2010). The USGS, in cooperation with the NMOSE, expanded and amended this monitoring network in 2010–11 to a total of 33 wells by including private stock and domestic wells for which access and permission could be obtained; the result is greater inclusion of eastern San Miguel County (fig. 5 and table 3). Twenty-four of these wells will be measured on a 5-year schedule to provide a record of background groundwater levels in San Miguel County.

Hydrographs (time series of water-level elevations) are presented in this report from six network monitoring wells measured on an approximately annual basis and having a period of record longer than 2 years as of October 2011, two of which are located in the Taylor well field, and from two network monitoring wells measured on a bimonthly basis and having 1 year of record as of October 2011 (fig. 1A and fig. 5), both of which are located south of the Taylor well field (fig. 1A and fig. 5). The water level measured in a well at a given point in time depends on water levels in the aquifer(s) in which the well was completed, as well as on the degree, if any, of hydraulic connection with adjacent aquifers. The water level in a well is also affected by how recently the well was pumped or by possible interference from pumping in nearby wells (Freeze and Cherry, 1979). Water-level measurements for which field observations indicated that the well was being pumped (P), the well was recently pumped (R), or a nearby well was pumping (S) are noted in the groundwater hydrographs presented. Water-level data from the USGS groundwater-monitoring network are archived in the USGS NWIS database and are available through the NWIS Web site (http://nwis.waterdata.usgs.gov/nm/nwis/gwlevels). Groundwater quality is summarized on the basis of Griggs and Hendrickson's (1951) characterizations; otherwise, no county-scale groundwater-quality data were located for review.

Griggs and Hendrickson's (1951) plate delineating physiographic areas and subareas of San Miguel County was scanned and imported into a geographic information system (GIS) project. The county map was registered to allow the boundaries of Griggs and Hendrickson's (1951) designated physiographic areas to be digitized as a basis for organizing hydrogeologic data for this report.

Data Gaps and Suggestions for Future Study

Information from the literature review and data identified from written reports and existing data sources were used to summarize the current understanding of the hydrologic resources of San Miguel County. Spatial data gaps and topical areas that might benefit from additional monitoring and/or assessment were identified, and suggestions for future study were presented.

Characterization of the Hydrologic Resources of San Miguel County

The current understanding of the hydrologic resources of San Miguel County is summarized in the following sections on the basis of information from the review of existing documents and data. Data are presented from streamgages and crest-stage gages currently (2011) operated in San Miguel County and are described with respect to monthly, annual, and daily discharge and with respect to gains and losses to local aquifers, as existing data permit. Surface-water quality is summarized with reference to U.S. Environmental Protection Agency (EPA) water-quality standards (U.S. Environmental Protection Agency, 2011a). Updates to the general stratigraphic and hydrostratigraphic framework and groundwater conditions of San Miguel County are presented. Hydrographs from monitoring wells are provided to establish recent (2011) trends in groundwater levels in parts of the county.

Surface-Water Resources

San Miguel County has been included in assessments of surface hydrologic resources at regional to drainage-basin scales (for example, Jansen, 1935; Dice, 1954; New Mexico State Engineer Office, 1975; Glorieta Geoscience, Inc., and James W. Siebert & Associates, 1990; Martinez, 1990; Chavez, 2004; and Aguirre, 2009). The most recent comprehensive study (Daniel B. Stephens & Associates, Inc., 2005) was part of the Region 8 Water Plan that included Mora, San Miguel, and Guadalupe Counties.

Streamflow characteristics in the Pecos River, Gallinas River, Canadian River, and their tributaries are discussed with reference to the location on the stream channel where data were collected by using the location as referenced in the name of the streamgage. Thus, a discussion of streamflow characteristics at the Pecos River near Pecos refers to streamflow characteristics of the river at the location of the streamgage near Pecos.

Table 3. Location of U.S. Geological Survey groundwater-monitoring wells in San Miguel County, New Mexico, period of record, measuring schedule, New Mexico Office of the State Engineer point of diversion number, and aquifer of completion.

[NAD 83, North American Datum of 1983; NMOSE, New Mexico Office of the State Engineer; POD, point of diversion; PROJECTED, extrapolated from surrounding Public Land Survey System]

Site identification number	Local identification number	Latitutude (NAD 83)	Longitude (NAD 83)	Period of record	Measuring schedule	NMOSE POD number	Aquifer of completion
353418105145601	16N.16E.33.143	35.5718	-105.2486	1999–2011	Semiannually	--	Santa Rosa Sandstone
352725105160901	14N.16E.08.321 (PROJECTED)	35.4568	-105.2691	2006–10	Annually	--	--
353001105402601	15N.12E.28.233 (PROJECTED)	35.5004	-105.6738	2006–10	Annually	--	--
354054105092101	Stock Well	35.6817	-105.1558	2006–10	Annually	--	--
354310105035801	Ranch Windmill	35.7194	-105.0662	2006–9	Annually	--	--
352949105144301	15N.16E.28.414	35.4970	-105.2454	2010–11	Recorder	--	--
353146105144801	15N.16E.16.1444	35.5295	-105.2467	2010–11	Recorder	--	--
353346105145201	15N.16E.04.242	35.5630	-105.2472	1999–2011	Semiannually	--	Santa Rosa Sandstone
353418105145601	16N.16E.33.143	35.5718	-105.2486		Semiannually	--	Santa Rosa Sandstone
351118104070501	11N.27E.09.333	35.1883	-104.1181	2011	5 year	TU-1056	--
351121105235001	11N.14E.12.44421	35.1888	-105.3972		5 year	UP-1912	Yeso Formation
351544104022101	12N.27E.14.434	35.2622	-104.0393	2011	5 year	TU-1084	--
351553103483301	12N.29E.13.423	35.2646	-103.8091	2011	5 year	--	--
351809104204101	12N.24E.01.211	35.3024	-104.3447	2011	5 year	CR-04906	--
352747104294301	14N.23E.03.3333	35.4629	-104.4952	2011	5 year	--	--
353136104401101	16N.21E.13.324	35.5267	-104.6698	2011	5 year	--	--
353211104455001	15N.20E.12.443	35.5364	-104.7640	2011	5 year	CR-04792	--
353301105154501	15N.16E.05.434 (PROJECTED)	35.5503	-105.2631		5 year	UP-2234	Dockum Group
353318105025601	15N.18E.05.431	35.5550	-105.0489	2011	5 year	UP-2836	--
353438105013701	16N.18E.33.423	35.5773	-103.0269	2011	5 year	UP-1399	--
353441105150401	16N.16E.33.114	35.5781	-105.2517		5 year	TAYLOR 5	Santa Rosa Sandstone
353643104055401	16N.27E.29.3	35.6119	-104.0984	2011	5 year	--	--
353657103535701	17N.28E.31.212	35.6158	-103.8988	2011	5 year	TV00200	--
353715105045601	16N.17E.16.4	35.6208	-105.0823	2011	5 year	--	--
353748104533801	16N.19E.01.324	35.6299	-104.8939	2011	5 year	CR-0-5102 POD1	--
353902104045601	16N.27E.04.123	35.6506	-104.0823	2011	5 year	TU-01599	--
353917104110801	17N.26E.31.44	35.6548	-104.1854	2011	5 year	--	--
354103103535701	17N.29E.19.343	35.6843	-103.8991	2011	5 year	TU-01015	--
354155104325101	17N.22E.13.442	35.6986	-104.5474	2011	5 year	CR-00282	--
354537104500301	18N.20E.29.421	35.7602	-104.8341	2011	5 year	CR-00706	--
354553104523901	18N.19E.25.231	35.7647	-104.8775	2011	5 year	CR-04267	--
354643104074101	10N.26E.24.2133	35.7787	-104.1274	2006–11	5 year	--	--
354724105144601	18N.16E.16.22331	35.7900	-105.2467		5 year	CR-2394	San Andres Limestone - Glorieta Sandstone

Streamflow in the Pecos River and Tributaries

The Pecos River and tributary stations include streamgages at the Rio Mora near Terrero, the Pecos River near Pecos, and the Pecos River near Anton Chico and a crest-stage gage at Tecolote Creek at Tecolote (fig. 2). Mean monthly discharge at the streamgages (fig. 4A–C) increases downstream and is generally bimodally distributed, with about 47 percent of annual streamflow occurring during April through June, corresponding to spring runoff, and with a lesser peak in August, representing about 11 percent of annual streamflow, corresponding to summer monsoonal rains. Annual discharge on the Rio Mora and the Pecos River generally increases downstream, with maximum annual discharge for the period of record on the Pecos River near Pecos and near Anton Chico occurring in 1941 and 1942, respectively (fig. 6A). There were periods when annual discharge near Anton Chico was of lower magnitude than near Pecos, indicating a loss of flow between the upstream Pecos River near Pecos and the downstream Pecos River near Anton Chico streamgages (fig. 6A) and possibly reflecting localized differences in runoff within the watershed, flow loss to groundwater, or irrigation diversions. Mean annual discharge for the 30-year period 1981–2010 was 26,330 acre-feet/year (acre-ft/yr) at Rio Mora near Terrero and 79,100 acre-ft/yr at Pecos River near Pecos (table 2). The Rio Mora enters the Pecos River about 5.6 mi upstream from the Pecos River near Pecos streamgage. Mean annual discharge at Pecos River near Anton Chico for the same period was 90,730 acre-ft/yr (table 2), an average gain of 11,630 acre-ft/yr over about 91 mi of river length.

In figure 7, the shape of the flow duration curve characterizes the hydrologic response of a basin and the expected distribution of flow in that basin on an annual basis. A curve with a steep slope through all portions of the curve denotes a stream with highly variable flow largely from direct runoff, whereas a curve with a flatter slope indicates the presence of surface-water or groundwater storage and a substantial component of base flow, which tends to stabilize flow through the year. The shape of the curve in the high-flow region (the left side of the curve) indicates the type of flood regime the basin is likely to have, whereas the shape of the curve in the low-flow region (the right side of the curve) indicates the ability of the basin to sustain low flows during dry seasons (Searcy, 1959; Oregon State University, 2011). A curve with a steep slope in the high-flow region reflects high-intensity, short-duration flows, such as might be expected in response to summer monsoonal rains. Snowmelt runoff or releases from reservoir storage for irrigation or flood control can result in sustained higher magnitude flows, resulting in a flatter curve near the high-flow region of the curve. In the low-flow region of the flow duration curve, an intermittent stream exhibits periods of no flow. A flatter slope in the low-flow

region indicates that moderate flows occur throughout the year because of either sustained base flow to the stream, reflecting a large groundwater storage capacity, or artificial streamflow regulation.

Days with zero discharge were included in the calculation of exceedance probability. The percent of time a given discharge was equaled or exceeded does not accumulate to 100 percent for the Pecos River near Anton Chico, Gallinas River near Colonias, and Canadian River near Sanchez curves because the zero values are not included in the logarithmic scale of the discharge axis but are included in the calculation of values (fig. 7A). The smallest discharge value recorded by the USGS was 0.01 ft³/s, below which discharge is considered to be zero. The difference between 100 percent and the maximum percentage value of a curve is the percentage of the discharge record with zero discharge (no flow).

Flow duration curves based on mean daily discharge values for the period 1981–2010 were computed for Rio Mora near Terrero, Pecos River near Pecos, and Pecos River near Anton Chico (fig. 7A). Excluding the upper and lower 5 percent of discharge values, where the curves change most rapidly, mean daily discharge for the middle 90 percent of discharge values ranges over two to three orders of magnitude, reflecting higher flows during spring and summer runoff as compared to flows during the winter months (fig. 4A–C). The middle 90 percent of expected mean daily discharge for the Rio Mora near Terrero ranges from 4.8 to 152 cubic feet per second (ft³/s), with a midvalue (50 percent probability of occurrence) of 15 ft³/s; for the Pecos River near Pecos, 20–430 ft³/s, with a midvalue of 51 ft³/s; and for the Pecos River near Anton Chico, 2.4–561 ft³/s, with a midvalue of 45 ft³/s (fig. 7A). The high-flow area of the curves are characterized by a short upward tail before the curve flattens, indicating that high-magnitude flows at these sites consist of a small contribution from high-intensity, short-duration monsoonal rains and a sustained snowmelt runoff component (fig. 7A), consistent with the distribution of the mean monthly discharge (fig. 4A–C). The low-flow area of the curves for the Rio Mora near Terrero and the Pecos River near Pecos are relatively flat compared to the curve for the Pecos River near Anton Chico and never drop to zero (fig. 7A), indicating sustained, perennial base flow. Although the Pecos River near Anton Chico is the most downstream gage and has the largest drainage area, low-flow values at that gage are lower than at the Rio Mora near Terrero or the Pecos River near Pecos. The slope of the curve for the Pecos River near Anton Chico is steeper than for the other two curves for low flows, with occasional days of zero flow recorded (29 days over the 30-year period considered). The decline in low-flow values at the downstream Pecos River near Anton Chico streamgage may reflect flow loss within the channel and withdrawals for irrigation between the streamgages at Pecos River near Pecos and Pecos River near Anton Chico.

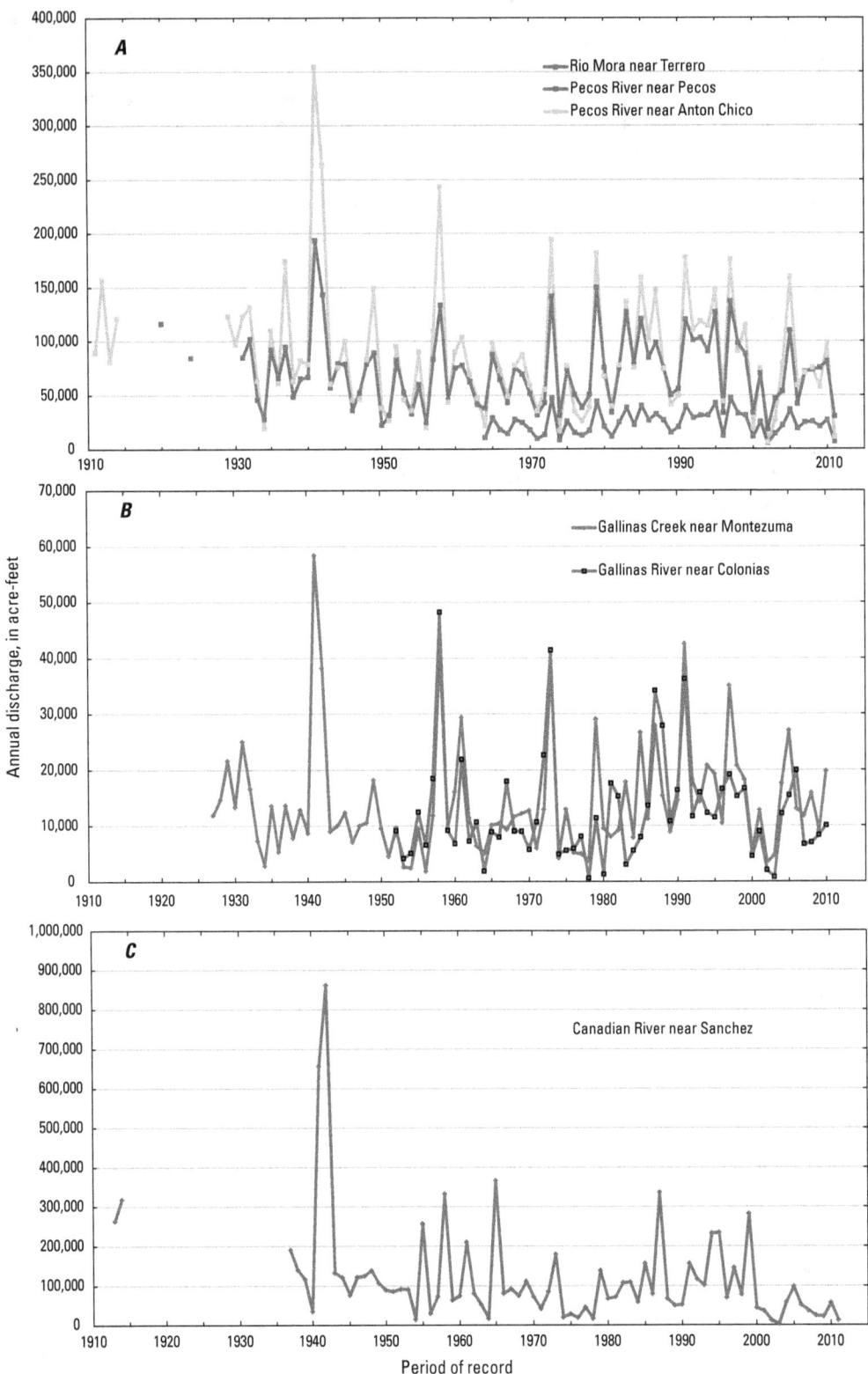

Figure 6.   Annual discharge for the period of record for six streamgages in San Miguel and Guadalupe Counties, New Mexico. *A*, Rio Mora near Terrero (08377900), Pecos River near Pecos (08378500), and Pecos River near Anton Chico (08379500). *B*, Gallinas Creek near Montezuma (08380500) and Gallinas River near Colonias (08382500). *C*, Canadian River near Sanchez (07221500).

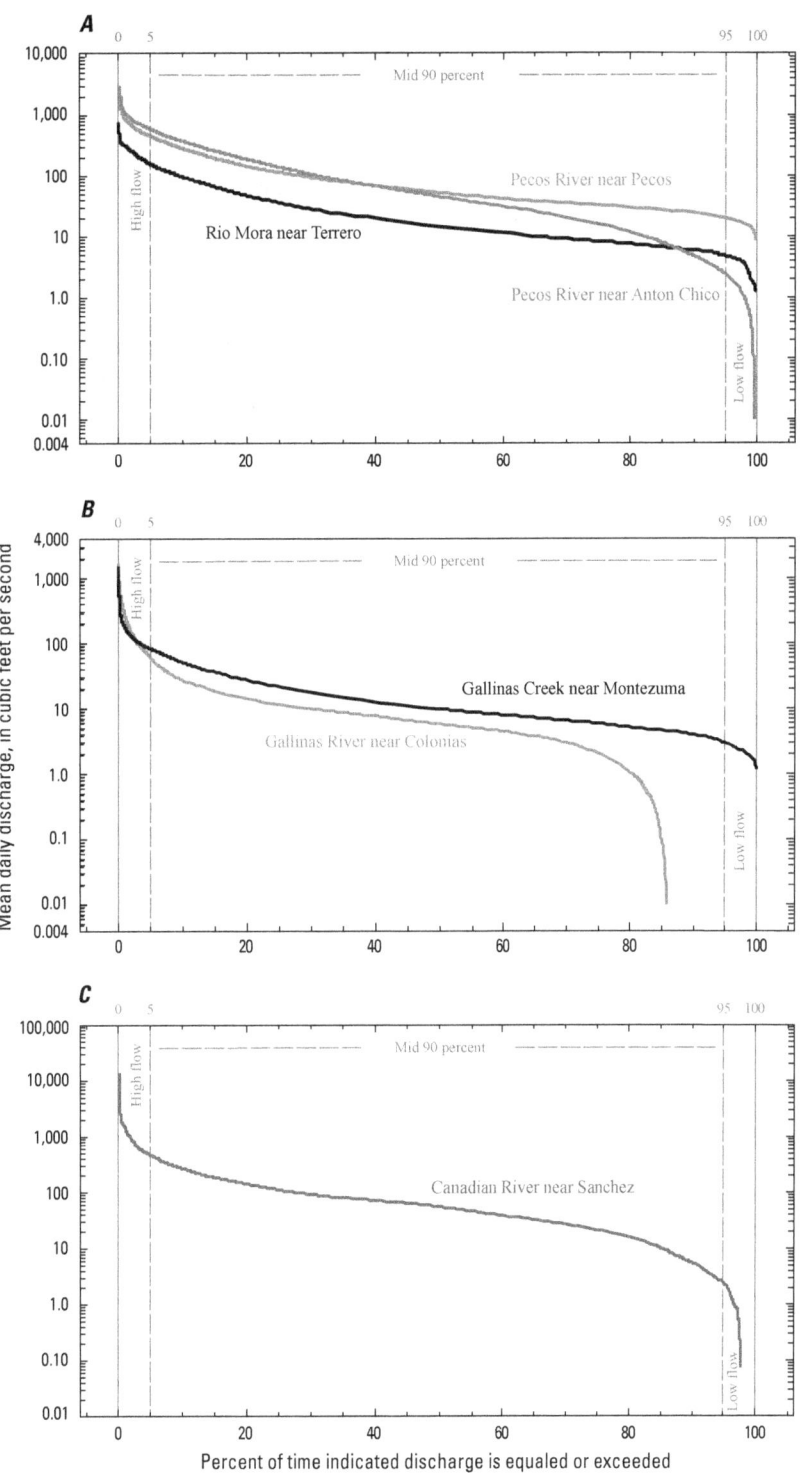

Figure 7.   Probability that a given discharge will be equaled or exceeded for the period 1981–2010 for six streamgages in San Miguel and Guadalupe Counties, New Mexico. *A*, Rio Mora near Terrero (08377900), Pecos River near Pecos (08378500), and Pecos River near Anton Chico (08379500). *B*, Gallinas Creek near Montezuma (08380500) and Gallinas River near Colonias (08382500). *C*, Canadian River near Sanchez (07221500).

The peak flow analysis describes the instantaneous annual maximum daily flow in terms of the probability of occurrence of a given magnitude flow in any given year on the basis of the probability of occurrence of that magnitude flow for the period of record at that gage (Flynn and others, 2006). For example, a discharge with a 2-year recurrence interval would have a 50 percent probability of occurrence in any given year, while a discharge with a 50-year recurrence interval would have a 2 percent probability of occurrence in any given year. Thus, the expected discharges for the 2- and 50-year recurrence interval flows indicate the range of expected streamflows for fairly common and more extreme flows at that location. Differences in the magnitude of peak discharges among gaging stations can be attributed to differences in such characteristics as size of the contributing drainage area, topography, land cover, antecedent soil moisture, or the areal extent and intensity of the precipitation event that resulted in the peak flow.

Observed instantaneous annual maximum peak discharges on the Pecos River and tributaries, represented by the 2- and 50-year recurrence interval flows, are, for the Rio Mora near Terrero, 257 and 910 ft³/s; for the Pecos River near Pecos, 623 and 2,622 ft³/s; for Tecolote Creek at Tecolote (Tecolote Creek is a tributary to the Pecos River with a similar drainage area to the Pecos River near Pecos [table 2]), 853 and 9,385 ft³/s; and for the Pecos River near Anton Chico, 6,029 and 24,250 ft³/s (table 4).

Streamflow in the Gallinas River and Tributaries

The Gallinas River and tributary stations include the continuous streamgages Gallinas Creek near Montezuma and Gallinas River near Colonias and the crest-stage gage Sandoval Canyon at Gallinas (fig. 2 and table 2). Mean monthly discharge on the Gallinas Creek near Montezuma is bimodally distributed, with 38 percent of the mean annual discharge in April and May, corresponding to spring runoff, and a lesser peak in August representing 14 percent of the mean annual discharge and corresponding to summer monsoonal rains (fig. 4D). Mean monthly discharge on the Gallinas River near Colonias shows a subdued response to spring runoff, with 16 percent of the mean annual discharge occurring in April and May and 47 percent in July and August corresponding to summer monsoonal rain (fig. 4E). Annual hydrographs at Gallinas Creek near Montezuma and Gallinas River near Colonias do not show a consistent increase or decrease between the two streamgages, and the hydrographs sometimes diverge in trend, for example, during the periods 1981–87, 1992–95, and 2007–8 (fig. 6B). Mean annual discharge for the period 1981–2010 was 16,260 acre-ft/yr at Gallinas Creek near Montezuma and 13,470 acre-ft/yr at Gallinas River near Colonias (table 2), a loss of about 2,800 acre-ft/yr along about 72 mi of river length.

Flow duration curves based on mean daily discharge values for the period 1981–2010 were computed for Gallinas Creek near Montezuma and Gallinas River near Colonias (fig. 7B). Excluding the upper and lower 5 percent of discharge values, where the curve changes most rapidly, mean daily discharge for the middle 90 percent of discharge values for Gallinas Creek near Montezuma ranges from 85 to 3.1 ft³/s, with a midvalue of 10 ft³/s. For the Gallinas River near Colonias, mean daily discharge ranges from 61 to 2.1 ft³/s between 5 and 75 percent probability of occurrence, after which the curve declines to zero, with a midvalue (50 percent probability of occurrence) of about 6 ft³/s. The low-flow area of the Gallinas Creek near Montezuma curve is relatively flat compared to the Gallinas River near Colonias curve, indicating more sustained base flow or snowmelt runoff than is indicated by the somewhat steeper slope of the Gallinas River near Colonias curve. These flow patterns are consistent with the distribution of mean monthly discharge (fig. 4D–E) and indicate that while discharge at Gallinas Creek near Montezuma is broadly distributed, discharge at Gallinas River near Colonias is dominated by summer monsoonal rains. The high-flow areas of the flow duration curves indicate a greater percentage of high-intensity, short-duration flow on the Gallinas than on the Pecos River (fig. 7A–B), as indicated by the comparatively longer vertical tails in the high-flow region of the curves. Comparison of the flatter slope of the Gallinas Creek near Montezuma curve to the Gallinas River near Colonias curve within the lower 5 percent of flow indicates that a more sustained snowmelt runoff component likely contributes to low flows on Gallinas Creek (fig. 7B). The low-flow portion of the Gallinas Creek near Montezuma curve indicates that, while discharge declines, flow remains perennial and has not been measured at less than 1.2 ft³/s at the streamgage during the period of record. By comparison, no flow was recorded at Gallinas River near Colonias about 14 percent of the time. The Gallinas River at this location is characterized by declining base flow or snowmelt runoff, with periodic high-flow events and periods of no flow. Observed instantaneous annual maximum peak discharges on the Gallinas River and tributaries, represented by the 2- and 50-year recurrence interval flows, are, for Sandoval Canyon at Gallinas, tributary to Gallinas Creek, 73 and 1,531 ft³/s; for Gallinas Creek near Montezuma, 514 and 6,118 ft³/s; and for the Gallinas River near Colonias, 2,851 and 15,040 ft³/s (table 4).

Streamflow in the Canadian River and Tributaries

The Canadian River and tributary stations include the streamgage at the Canadian River near Sanchez and crest-stage gages at Lagartija Creek Tributary near Sanchez, Trementina Creek at Trementina, and Garita Creek Tributary near Variadero. Lagartija Creek is tributary to the Canadian River, and Trementina Creek and Garita Creek Tributary flow into Conchas Lake. Mean monthly discharge at the Canadian River near Sanchez is bimodally distributed, with about 29 percent of flow in May through June, corresponding to spring runoff, and 15 percent in August, corresponding to summer monsoonal rains (fig. 4F). As for the other three

Table 4. Annual maximum peak discharges at selected recurrence intervals for streamgages and crest-stage gages in San Miguel and Guadalupe Counties, New Mexico.

[USGS, U.S. Geological Survey]

USGS station identification number	Name	Recurrence interval (years)	Annual maximum peak discharge (cubic feet per second)
08377900	Rio Mora near Terrero	2.0	257
		10	565
		50	910
08378500	Pecos River near Pecos	2.0	623
		10	1,550
		50	2,622
08379300	Tecolote Creek at Tecolote	2.0	853
		10	3,887
		50	9,385
08379500	Pecos River near Anton Chico (Guadalupe County)	2.0	6,029
		10	14,840
		50	24,250
08380300	Sandoval Canyon at Gallinas	2.0	73
		10	451
		50	1,531
08380500	Gallinas Creek near Montezuma	2.0	514
		10	2,360
		50	6,118
08382500	Gallinas River near Colonias (Guadalupe County)	2.0	2,851
		10	8,796
		50	15,040
07221500	Canadian River near Sanchez	2.0	5,996
		10	27,340
		50	74,820
07221600	Lagartija Creek Tributary near Sanchez	2.0	207
		10	665
		50	1,454
07222300	Trementina Creek at Trementina	2.0	1,639
		10	6,929
		50	15,570
07222800	Garita Creek Tributary near Variadero	2.0	1,995
		10	3,567
		50	9,643

streamgages which cover a similar time period (fig. 6*A–B*), the maximum annual discharges for the period of record for Canadian River near Sanchez occurred in 1941 and 1942 (fig. 6*C*). Mean annual discharge for the 30-year period 1981–2010 was 97,800 acre-ft/yr (table 2).

A flow duration curve based on mean daily discharge values for the period 1981–2010 was computed for the Canadian River near Sanchez (fig. 7*C*). Excluding the upper and lower 5 percent of discharge values, where the curve changes most rapidly, mean daily discharge for the middle 90 percent of discharge values for the Canadian River near Sanchez ranges from 505 to 2 ft³/s, after which the curve declines to zero, with a midvalue (50 percent probability of occurrence) of 54 ft³/s. The curve in the high-flow area is constrained by a few high-intensity, short-duration events over the period considered, and then the slope of the curve declines steadily, with a sharp drop at the low-flow portion of the curve. The Canadian River near Sanchez drains 5,712 mi² (table 2) of the upper Canadian River basin (fig. 2) and is dry at this location about 2 percent of the time. The gradual decline of the flow duration curve in the middle 90 percent likely reflects greater temporal and spatial variability with respect to precipitation, runoff, and potential transmission losses associated with the larger and more diverse contributing source area, as compared to the contributing source areas for the Pecos and Gallinas Rivers and tributaries (fig. 7*A–F*). Observed instantaneous annual maximum peak discharges on the Canadian River and tributaries, represented by the 2- and 50-year recurrence interval flows, are, for Lagartija Creek Tributary near Sanchez, 207 and 1,454 ft³/s; for Trementina Creek at Trementina, 1,639 and 15,570 ft³/s; for Garita Creek Tributary near Variadero, 1,995 and 9,643 ft³/s; and for the Canadian River near Sanchez, 5,996 and 74,820 ft³/s (table 4).

Major Surface-Water Bodies and Springs

There are three major lakes in San Miguel County: Lake Isabel, Conchas Lake on the Canadian River, and Storrie Lake on the Sapello River (New Mexico State Engineer Office, 1975; Daniel B. Stephens & Associates, Inc., 2005). Conchas Lake is a 25-mi-long water body with a surface area of about 6,419 acres and average storage of 61,532 acre-feet (acre-ft) (Daniel B. Stephens & Associates, Inc., 2005) that is fed by the Canadian and Conchas Rivers; the reservoir supplies appropriated water to irrigated lands around Tucumcari, about 35 mi to the southeast (U.S. Bureau of Reclamation, 1979). Lake Isabel, averaging 530 acres in surface area and 6,500 acre-ft of storage (Daniel B. Stephens & Associates, Inc., 2005), also supplies irrigation water. Storrie Lake, averaging 907 acres in surface area and 21,747 acre-ft of storage (Daniel B. Stephens & Associates, Inc., 2005), supplies water to Las Vegas and acequias of the Las Vegas Acequia Association (Ebright, 2009).

Twenty-two springs have been located in San Miguel County (Griggs and Hendrickson, 1951; White and Kues,

1992). Most are reported to yield less than 12 gallons per minute, with five of the springs reported to yield 40–150 gallons per minute and one reported to yield an anomalously high 400 gallons per minute. No other published sources updating spring yields reported by Griggs and Hendrickson (1951) were located.

Surface-Water Budget and Potential for Groundwater Recharge from Streamflow

Longworth and others (2008) reported an annual total for 2005 of 71,152 acre-ft of surface-water withdrawals in San Miguel County, including 32,292 acre-ft lost to reservoir evaporation. Of the total surface-water withdrawals, 2,314 acre-ft were allocated to public water supply, 277 acre-ft to livestock, and 164 acre-ft to commercial use. In 2005, an annual total of 36,105 acre-ft of surface-water withdrawal supplied 10,986 irrigated acres for agriculture (Longworth and others, 2008). In a water budget for river reaches in San Miguel County, based on median conditions for the period 1950–2002, Daniel B. Stephens & Associates, Inc. (2005), reported 10,783 acre-ft/yr in surface-water depletions on the Pecos River between the Pecos River near Pecos and the Pecos River near Anton Chico, and an estimated 4,351 acre-ft/yr gain due to irrigation return flow and a gain of 10,038 acre-ft/yr attributed to other sources. On the Gallinas River between Montezuma and Colonias, surface-water withdrawals totaling 17,462 acre-ft/yr were attributed to use for public supply and irrigation and to losses from reservoir and riparian evapotranspiration; 4,031 acre-ft/yr of gain in the water balance was attributed to return flow; and 9,655 acre-ft/yr of gain was attributed to other sources, based on median conditions for the period 1950–2002 (Daniel B. Stephens & Associates, Inc., 2005). The water balance for this reach of the Gallinas River implies a gain in streamflow despite the loss calculated solely on the difference between median discharge at the upstream and downstream gages (discussed in the section "Streamflow in the Gallinas River and Tributaries" and based on mean annual discharge for the period 1981–2010). A surface-water withdrawal of 3,414 acre-ft/yr for irrigation occurred in San Miguel County above the Canadian River near Sanchez (Daniel B. Stephens & Associates, Inc., 2005).

A stream channel may exchange water with the aquifer that it overlies. In a gaining reach, there is a gain in flow to the stream channel, fed by a corresponding loss from groundwater; in a losing reach, there is a loss in flow from the stream channel and a corresponding gain to the underlying aquifer. A stream may be alternately gaining and losing along the length of the channel (Rushton, 2007). Focused groundwater recharge, streamflow transmission loss that recharges the underlying groundwater system, is characteristic of many arid and semiarid streams (Stonestrom and others, 2007). Both focused groundwater recharge and the groundwater contribution to streamflow along a stream reach may be estimated by a seepage survey, which is a synoptic measurement of discharge at the upstream and downstream ends of a reach and of all identified inflows and outflows within the reach.

Seepage surveys conducted in the upper Gallinas River reported the presence of both losing and gaining reaches in that part of the river (Jansen, 1935; U.S. Geological Survey, 1978, 2006), but the seepage investigations of the Gallinas River were not areally extensive. No seepage surveys were identified for other reaches of the Gallinas River or for other rivers in San Miguel County. The geologic strata through which the rivers of San Miguel County flow vary with location, and the degree of hydrologic connection with the underlying aquifers is not known in detail; as a result, extrapolation of results from the available seepage surveys to other parts of the river network is not feasible. In the Region 8 Water Plan, Daniel B. Stephens &Associates, Inc. (2005), developed groundwater budgets for declared underground water basins in the three-county study area but did not develop groundwater budgets at a more detailed scale because of the lack of local-scale supporting data. Similarly, supporting data are not sufficient to account for potential recharge to underlying aquifers from streams in San Miguel County, making a county-wide coupled surface and subsurface water balance infeasible.

Surface-Water Quality

A baseline water-quality study of the headwaters of Gallinas Creek upstream from the Las Vegas water-supply diversion, located about 0.5 mi downstream from the Gallinas Creek near Montezuma streamgage, was conducted from 1987 to 1990 (Garn and Jacobi, 1996). The study found water quality in the upper reaches of Gallinas Creek to be nonimpaired and in the lower reaches to be slightly impaired with respect to pH, temperature, and turbidity when compared to New Mexico State water-quality standards. Water-quality concerns included grazing, recreational use, septic systems along the river, roads, and insect invasion in national forest lands (New Mexico Soil Conservation Service, 1994). Sampling for major ions, nutrients, and field parameters (temperature, dissolved oxygen, pH, and specific conductance) by the New Mexico Environment Department (NMED) in 2007–9 to assess the potential impacts of forest thinning in the Gallinas River watershed upstream from Las Vegas found exceedances of only the temperature and turbidity criteria. The temperature exceedance for the High Quality Aquatic Life designated use is consistent with the historical finding of nonsupport of that use in this assessment unit (New Mexico Environment Department, 2009, 2010).

Total maximum daily load (TMDL) studies assess whether a water body is in compliance with water-quality standards, address aggregate pollutant loads from point and nonpoint sources, and identify the amount or concentration of a given pollutant that the water body under consideration can contain while allowing the water body to remain viable for its designated use(s) (New Mexico Environment Department, 2011). Results of TMDL studies were summarized in the New Mexico water-quality assessment report (U.S. Environmental Protection Agency, 2011a) for the Pecos Headwaters watershed, which includes the Pecos and Gallinas Rivers, and the upper Canadian watershed. Within San Miguel County, approximately 270 river miles of the Pecos Headwaters watershed and 100 river miles of the upper Canadian watershed were listed as impaired, meaning that water quality for specific criteria did not meet standards for a specific use in monitored reaches. Impairments were primarily for the uses High Quality Cold Water Aquatic Life and Marginal Cold Water Fisheries but included the uses Secondary Contact Recreation and Domestic Water Supply in selected reaches (U.S. Environmental Protection Agency, 2011a). Criteria for which there were findings of impairment were primarily temperature and turbidity but also included nutrients, ammonia, nitrate and nitrite, pH, dissolved oxygen, specific conductance, siltation, and mercury in fish tissue in selected reaches. Causes of impairment were variously attributed to loss of riparian habitat, rangeland grazing, watershed runoff following forest fires, road runoff, streambank modification or destabilization, livestock, low flow alterations, *Escherichia coli*, aquaculture (permitted), atmospheric deposition, inactive mine reclamation, and natural sources.

Conchas Lake is impaired as a warm water fishery because of eutrophication and the presence of mercury and polychlorinated biphenyls (PCBs) in fish tissue exceeding established criteria. Storrie Lake is impaired for cold and warm water fisheries because of the presence of mercury in fish tissue exceeding established criteria. McAllister Lake is impaired for cold water fisheries because of atmospheric deposition of arsenic (U.S. Environmental Protection Agency, 2011a).

Three sites on the upper Pecos River, all part of an inactive lead-zinc mining and milling complex near the town of Pecos (Robinson, 1995), are listed as Comprehensive Environmental Response, Compensation, and Liability Information System (CERCLIS) sites under the EPA Superfund program: Terrero Mine (EPA identification [ID] NMD986668820) near Terrero, El Molino Mill (EPA ID NMD981057292) near Pecos, and East Pecos (EPA ID NM0000605422) near Pecos (fig. 1*A*). Leachate from mine waste was cited as a concern by Johnson and Deeds (1995a, b). Terrero Mine has undergone remediation and is listed as "archived" in the EPA Abandoned Mine Lands CERCLIS inventory (U.S. Environmental Protection Agency, 2011b), meaning that the site does not require cleanup under the Federal Superfund program. El Molino Mill is listed as a No Further Remedial Action Planned (NFRAP) site (U.S. Environmental Protection Agency, 2011c), and the East Pecos site is listed as Preliminary Assessment Start Needed (U.S. Environmental Protection Agency, 2011d).

Groundwater Resources

The first county-wide characterization of the geology and groundwater resources of San Miguel County was compiled in Griggs and Hendrickson (1951). The general stratigraphy and hydrogeologic framework of San Miguel County are described here on the basis of Griggs and Hendrickson (1951), updated with data compiled from the NMWRRS and USGS NWIS Groundwater Site Inventory databases (as summarized in Matherne and Stewart, 2012) and with reports that provide localized hydrogeologic information. Groundwater resources of San Miguel County are summarized and updated in this report by using the physiographic provinces described in Griggs and Hendrickson (1951) (fig. 1*B*) as an organizational framework.

General Stratigraphic and Hydrogeologic Framework

The availability of groundwater is a function of the configuration of water-bearing rocks and unconsolidated sediments in the subsurface, known as the hydrogeologic framework, and of groundwater recharge to, discharge from, and water flux through that framework (Heath, 1983).

The water-bearing rock strata and important geological and structural features of San Miguel County were reported at a regional scale by Griggs and Hendrickson (1951) and at a variety of local to regional scales by others (for example, Read and others, 1944; Spiegel, S.J., 1956; Baltz, 1972; Glorieta Geoscience, Inc., and James W. Siebert & Associates, 1990; Daniel B. Stephens & Associates, Inc., 2005). Hydrostratigraphic nomenclature and description of rock-aquifer units are summarized in time-stratigraphic order in table 5, where nomenclature and age ranges correspond to cited reports. The USGS National Geologic Map Database Geologic Names Lexicon (GEOLEX; U.S. Geological Survey, 2011b) was the authority for accepted rock-formation nomenclature, except where noted. Reported rock-strata thickness estimates may not be equivalent to aquifer thicknesses, particularly with respect to crystalline bedrock, for which fracture flow may be dominant (Freeze and Cherry, 1979).

The oldest described water-bearing unit in San Miguel County is Precambrian granite (Griggs and Hendrickson, 1951). Precambrian metamorphic and igneous rocks—which include gneiss, schist, quartzite, and pegmatite, as well as granite—form the cores of two southern Rocky Mountain ridges that bound the upper Pecos River Valley (Griggs and Hendrickson, 1951, p. 22; Baltz, 1972). The Precambrian Embudo(?) granite of Montgomery (the formation name, with question mark, is as noted by Baltz [1972]; this usage is not included in GEOLEX) was mapped by Baltz (1972) in the upper reaches of Gallinas Creek. The Tusas Mountain Granite is the only Precambrian granitic aquifer identified by name in NWIS records of wells located in San Miguel County

(Matherne and Stewart, 2012); no other Precambrian granitic aquifers are identified (table 5). Granite may bear groundwater in faults and fractures at shallow depths (Freeze and Cherry, 1979); however, with increasing depth and associated increasing overburden weight, the volume of water-bearing fractures and associated groundwater in such crystalline rock decreases markedly (Freeze and Cherry, 1979). Fractured Precambrian granite is considered to be the basal (bedrock) water-bearing unit in San Miguel County and either crops out at land surface or is overlain by Paleozoic sedimentary rocks (table 5). In some areas, limestones of the Mississippian–Devonian Arroyo Peñasco Group, including the Espiritu Santo Formation (Baltz, 1972; Glorieta Geoscience, Inc., 1986) and/or the Mississippian Tererro[1] Formation (Baltz, 1972; Glorieta Geoscience, Inc., 1986; Garn and Jacobi, 1996), unconformably overlie granitic bedrock (table 5). Baltz (1972) mapped these rocks in the northwest quadrant of the Gallinas Creek area (fig. 1*B*).

Unconformably overlying either granitic bedrock or limestone of the Arroyo Peñasco Group is the Paleozoic–Mississippian/Pennsylvanian Sandia Formation, which consists of interbedded limestone, shale, and sandstone (table 5). Overlying the Sandia Formation is the Madera Group, consisting of the Porvenir and Alamitos Formations. In combination, these rocks range from not present to as much as 2,000 ft in thickness (Griggs and Hendrickson, 1951); Baltz (1972) described local thicknesses in the same range in the Gallinas Creek area. Griggs and Hendrickson (1951) used the term "Magdalena Group" to describe these Pennsylvanian rock formations. Although this terminology was formally discontinued in 1984 (Baltz and Myers, 1984; MJDarrconsult, Inc., 2003), it persists in San Miguel County hydrologic reports, likely attributable to ongoing reliance on Griggs and Hendrickson's (1951) report.

Unconformably overlying Precambrian or older Paleozoic rocks are Pennsylvanian to Permian interbedded sandstones, shale, and limestone of the Sangre de Cristo Formation (table 5). In San Miguel County, the Sangre de Cristo Formation ranges between 600 and 1,300 ft in thickness (Griggs and Hendrickson, 1951; Baltz, 1972). The upper Sangre de Cristo Formation grades into the distinctive orange to orange-red Permian Yeso Formation (table 5) (Griggs and Hendrickson, 1951), which is composed of interbedded siltstone and sandstone and ranges from 170 to 1,000 ft in thickness (Baltz, 1972; Griggs and Hendrickson, 1951). The Permian Glorieta Sandstone overlies the Yeso Formation. The Glorieta Sandstone ranges from 100 to 240 ft in thickness (Griggs and Hendrickson, 1951; Baltz, 1972) (table 5). Griggs and Hendrickson (1951) placed the Glorieta Sandstone at the base of the San Andres Formation and noted that the Glorieta Sandstone contains a limestone middle member that may bear water and, where present, be as much as 30 ft thick.

Baltz (1972) indicated that the Bernal Formation overlies the Glorieta Sandstone and further noted that the lower part

_____

[1]Alternate spelling of "Terrero" was observed in the literature for this unit.

of the Bernal Formation may be equivalent to the San Andres Formation. The formal usage of the name "Bernal Formation" has since been discontinued in favor of the name "Artesia Formation" (U.S. Geological Survey, 2011b). For clarity, these rocks are referred to informally herein as the "Artesia (Bernal) Formation" (table 5). GEOLEX indicates that rocks formerly assigned to the Bernal Formation overlie the San Andres Formation (U.S. Geological Survey, 2011b) in the far western part of San Miguel County; Glorieta Geoscience, Inc. (1997a), also identified the occurrence of the Bernal Formation and placed these rocks above the San Andres Formation. Artesia (Bernal) Formation rocks have been correlated to members of the Permian Artesia Group in southern parts of San Miguel County and to the Anton Chico Member of the Triassic Moenkopi Formation in northern parts of San Miguel County (Lucas and Hayden, 1991; U.S. Geological Survey, 2011b). The Artesia (Bernal) Formation crops out along El Creston hogback (Baltz, 1972), the hogback ridge just west of Las Vegas. Albright (1962), in his study of the Taylor well field near Las Vegas (fig. 1A), noted that groundwater borne in the Artesia (Bernal) Formation is heavily mineralized and attributed degraded water quality in one production well in the Taylor well field to partial completion in this formation. Lazarus and Drakos (1998) described the Artesia (Bernal) Formation as an aquitard in the vicinity of the Taylor well field, southwest of Las Vegas, indicating the large variability of hydrologic characteristics of this formation in San Miguel County.

Unconformably overlying Paleozoic and/or Artesia (Bernal) Formation rocks are sedimentary rocks of the Dockum Group of the Mesozoic Triassic period (Baltz, 1972). The stratigraphic designations and associated nomenclature of these rocks are in revision (Lucas and Hunt, 1987, 1989; see table 5), but usage of the Dockum/Chinle terminology persists in hydrologic reports of San Miguel County written since 1987. The hydrostratigraphy in table 5 follows Griggs and Hendrickson's (1951) stratigraphic designations and naming conventions, since nomenclature changes are not yet settled (2011). Dockum Group rocks are composed of the basal Santa Rosa Sandstone, which is overlain by the Chinle Formation (Griggs and Hendrickson, 1951). In some locations, the Santa Rosa Sandstone is interbedded with as much as 100 ft of shale (Griggs and Hendrickson, 1951, p. 26). The Chinle Formation consists of two shale members which, in some locations, may contain water-bearing sandstone lenses. The shale members are interbedded with a water-bearing sandstone member that ranges between 45 and 165 ft in thickness and in places is interbedded with shale. Griggs and Hendrickson (1951, p. 25) note that Dockum Group rocks range in combined thickness from about 1,000 to 1,200 ft. Aquifers contained within Dockum Group water-bearing rocks are of variable but lesser thicknesses (Griggs and Hendrickson, 1951).

Where present, the Jurassic Entrada Sandstone of the San Rafael Group unconformably overlies Triassic rocks

(table 5) (Baltz, 1972). The Entrada Sandstone ranges from "inconspicuous in the hogback zone west of Las Vegas" (Griggs and Hendrickson, 1951, p. 27) to about 120 ft thick (Baltz, 1972) and is overlain, if present, by up to 25 ft (Baltz, 1972) of the Todilto Limestone Member of the Wanakah Formation of the San Rafael Group (U.S. Geological Survey, 2011b). The Todilto Limestone is overlain by the Morrison Formation, which ranges in thickness from 250 ft (Griggs and Hendrickson, 1951) to as much as 445 ft (Baltz, 1972). Griggs and Hendrickson (1951) noted that these Jurassic rocks were not known to contain highly productive aquifers in San Miguel County.

Cretaceous rocks unconformably overlie Jurassic rocks (table 5) (Baltz, 1972). The base of the Cretaceous section is composed of the Dakota Sandstone. The Purgatoire Formation is present below the Dakota Sandstone in areas northeast of San Miguel County but is not observed within the county (Kilmer, 1987; U.S. Geological Survey, 2011b). The Dakota Sandstone ranges in thickness from 100 to 250 ft (Baltz, 1972; Griggs and Hendrickson, 1951). Overlying the Dakota Sandstone are members of the Upper Cretaceous Mancos Shale (King, 1974); from older to younger the sequence consists of the conformable Graneros Member, the Greenhorn Limestone Member, the Carlile Member, and the Niobrara Member (table 5). North of San Miguel County these rocks have been assigned membership in the Colorado Group (U.S. Geological Survey, 2011b; M.J. Darr, oral commun., 2011). In the Gallinas Creek area of San Miguel County, the Graneros Member ranges in thickness from 215 to 250 ft, the Greenhorn Limestone Member is 40–60 ft thick, the Carlile Member is about 340 ft thick, and the Niobrara Member varies from 250 to 700 ft thick (Baltz, 1972). The Niobrara and Greenhorn Members are currently the focus of regional oil and gas exploration, depending upon the organic fractions of the underlying Carlile and Graneros Members, respectively (Durham, 2011). The quality of groundwater contained in these rocks may be affected if the rocks also contain oil and/or gas.

Although the Ogallala Formation of Cenozoic Neogene age is found locally in San Miguel County in thicknesses of as much as 50 ft, it is not regionally extensive and, in general, is not present because of erosion (Griggs and Hendrickson, 1951). The Tesuque Formation of the Santa Fe Group, also of Cenozoic Neogene age, was identified in the far western part of San Miguel County (Glorieta Geoscience, Inc., 1997a) in the Northern Rio Grande underground water basin (fig. 5).

Unconsolidated alluvium and pediment gravel and some igneous (volcanic) rocks of Quaternary age were described by Griggs and Hendrickson (1951). Alluvium, deposited by streamflow along perennial, intermittent, or ephemeral stream channels, may bear water sufficient for domestic purposes, for stock watering, or for small-scale irrigation projects (Griggs and Hendrickson, 1951).

Table 5.   Summary of hydrostratigraphic nomenclature and description of rock-aquifer units in San Miguel County, New Mexico, as derived from cited literature.

[ft, foot; ft²/d, square foot per day; ft/d, foot per day; ft³/ft³, cubic foot per cubic foot; NA, not applicable; --, no data]

Era[1,2,3]	Period[1,2,3] (dashed border indicates uncertainty or inconsistency between sources)	Epoch/ series/ other temporal subdivision	Group[1]	Formation or member[1,2,3]	Lithology[2,3]	Thickness[2,3] (ft)
Cenozoic	Quaternary	NA	NA	Alluvium	Alluvium; unconsolidated deposits of sediment including pediment, terrace, and other deposits of gravel, silt, sand, and locally, caliche	Multiple
	Neogene	Miocene and Pliocene	NA	Ogallala Formation	Unconsolidated to consolidated pebbles, gravel, sand, silt, and clay with caliche	0–50
		Miocene and Pliocene	Santa Fe Group (far west parts of San Miguel County)	Tesuque Formation	Pinkish to tan ledge-forming soft sandstone	--
Mesozoic	Cretaceous	Late	NA	Niobrara Member of Mancos Shale (b)	Gray shale interbedded with siltstone and a few thin limestone beds	250–700
				Carlile Member of the Mancos Shale	Gray shale interbedded with silstone and a few thin limestone beds	340
				Greenhorn Limestone Member of the Mancos Shale	Light gray limestone interbedded with gray-calcareous shale	40–60
				Graneros Member of the Mancos Shale	Dark gray shale, a few bentonite beds	215–250
		Early	NA	Dakota Sandstone	Brown sandstone interbedded with gray shale	100–250
	Jurassic	Late	San Rafael Group	Morrison Formation	Interbedded greenish-gray siltstone, shale, sandstone, and conglomeritic sandstone	250–445
		Middle		Todilto Limestone Member of the Wanakah Formation	Limestone, slightly sandy and gypseous	0–25
				Entrada Sandstone	Buff sandstone	0–120
	Triassic	Late	Dockum Group (nomen- clature in revision) (a)	Chinle Formation (forma- tion versus group status in revision) (a)	Interbedded shale and sandstone, limestone lenses, limestone pebbles and limestone pebble conglomer- ate, water-bearing sandstone middle member	750–1,150 (middle member: 45–165)
				Santa Rosa Sandstone (Santa Rosa Formation)	Light gray, tan, brown, gray sandstone interbedded with shale; limestone and chert pebble inclusions	200–400
		Middle		Anton Chico Member of the Moenkopi Formation[1,7,16] (depending on location may be equivalent to Artesia (Bernal) Formation[1,16])		[16]72 (at type section)

Suitability as aquifer[2, or as noted]	Transmissivity (ft²/d) (c)	Hydraulic conductivity (ft/d) (c)	Storage coefficient (ft³/ft³, unless noted otherwise)
Suitable	[14]74–1,685 (alluvium and weathered bedrock, undifferentiated)	--	--
Locally useful but not extensive in San Miguel County	--	--	--
--	[7]490	--	Specific yield: [25]0.04–0.05
Locally suitable[2]; regional exploration for oil and gas[21]	--	--	--
Locally suitable[2, 18]	--	--	--
Locally suitable[2]; regional exploration for oil and gas[21]	--	--	--
Locally suitable[2]	--	--	--
Regionally suitable except near escarpments[2, 18]	--	--	--
Not known to contain productive aquifers[2, 18]; Entrada Sandstone may be a viable exploration target[24]	--	--	--
	--	--	--
	--	--	--
Suitable	[19]227–2,273 Conchas Lake vicinity: [17]188–194 Glorieta - Santa Rosa undifferentiated (Taylor well field): [10, 15]570–4,300 Glorieta - Santa Rosa - Chinle - Bernal undifferentiated (Taylor well field): [20]470–790	[19]0.76–7.6	Glorieta - Santa Rosa undifferentiated (Taylor well field): [10, 15]1.5E-4; 3E-3 to [20]3E-5
--	--	--	--

Table 5.   Summary of hydrostratigraphic nomenclature and description of rock-aquifer units in San Miguel County, New Mexico, as derived from cited literature.—Continued

[ft, foot; ft²/d, square foot per day; ft/d, foot per day; ft³/ft³, cubic foot per cubic foot; NA, not applicable; --, no data]

Era[1,2,3]	Period[1,2,3] (dashed border indicates uncertainty or inconsistency between sources)	Epoch/ series/ other temporal subdivision	Group[1]	Formation or member[1,2,3]	Lithology[2,3]	Thickness[2,3] (ft)
Paleozoic	Permian	Late	Artesia Group	Artesia (Bernal) Formation (in southeast New Mexico equivalent to Grayburg and Queen Formations of Artesia Group)[16]	Purplish-red sandstone interbedded with siltstone; some local gypsum beds	115–140
		Early	NA	San Andres Limestone, San Andres Formation	Gray limestone, sandy	0–180
				Glorieta Sandstone Member of the San Andres Formation (middle unit is water bearing)	Yellow to buff orthoquartzitic sandstone with thin shale beds Middle unit bears water	100–240 (30, middle water-bearing unit)
				Yeso Formation	Orange-red sandstone, siltstone, and shale, interbedded with a few thin limestone beds	170–1,000
	Pennsylvanian/ Permian	Temporal divisions uncertain		Sangre de Cristo Formation	Red, purple, and greenish-gray shale interbedded with arokosic sandstone; contains thin beds of nodular or gray limestone; basal bed is massive arkosic conglomerate	600–1,300
			Madera Group	Alamitos Formation	Red, gray, and greenish-gray shale interbedded with gray limestone; nodular limestone; red marly shale at base	0–1,000
	Pennsylvanian	Middle		Porvenir Formation	Gray fossiliferous limestone, interbedded with dark gray shale and sandstone	0–1,200
		Early	NA	Sandia Formation	Interbedded limestone, shale, and sandstone	0–400
	Mississippian	Late	Arroyo Peñasco Group	Tererro Formation	Sandy clastic limestone; gray crystalline limestone	0–100
	Devonian	Early		Espiritu Santo Formation	Dark gray limestone, dolomitic limestone, sandy limestone, basal sandstone	
Proterozoic (Eon)	Precambrian	NA	NA	Tusas Mountain Granite	Granite	--
				Embudo(?) Granite (of Montgomery)[3]	Granite and/or granodiorite intruded into schistosic rocks	--
				Pegmatites, Igneous and metamorphic bedrock	Pegmatite dikes, gneiss and schist inclusions; schist, gneiss, metaquartzite	--

Oldest ←

(a) Proposed nomenclature change to (younger to older rocks): Redonda Formatio; Bull Canyon Formation; Trujillo Formation; Garita Creek Formation [1, 23]

(b) Oil and gas exploration ongoing [21]

(c) Transmissivity values reported in other units converted to ft²/d Hydraulic conductivity values reported in other units converted to ft/d

Sources:

[1]U S Geological Survey, 2011b [GEOLEX, National Geologic Names Lexicon]

[2]Griggs, R L , and Hendrickson, G E , 1951

[3]Baltz, E H , 1972

[4]Baltz, E H , and Myers, D A , 1984

[5]John Shomaker & Associates, 2007

[6]Souder, Miller & Associates, 1999

[7]Glorieta Geoscience, Inc , 1997a

[8]MJDarrconsult, Inc , 2003

[9]Energia Total, Ltd , 1997

[10]Lazarus, J , and Drakos, P G , 1998

[11]Corbin Consulting, Inc , 2006

[12]Glorieta Geoscience, Inc , 2003

Suitability as aquifer[2, or as noted]	Transmissivity (ft²/d) (c)	Hydraulic conductivity (ft/d) (c)	Storage coefficient (ft³/ft³, unless noted otherwise)
Very poor quality[22], aquitard in some areas[10]	--	--	--
	120,000	--	--
--	[5]690; [19]0 13–9,400; Glorieta - Santa Rosa undifferentiated: [10, 15]570–4,300; [17]188–194	--	Specific yield: [11]0.01–0.1 [10, 15]1.5E-4
Water may be mineralized[2]; water quality is poor[18]	Yeso/Sangre de Cristo undifferentiated: [5]985–5,300	--	--
Suitable	[7, 12]0.5–12; [9]295–1,062; [12]4–23; [13]13–20 Yeso/Sangre de Cristo undifferentiated: [5]985–5,300	--	[13]1.85 E-4
Suitable	Madera: [13]35–43; [6]43; [7]35–43 Madera/Sangre de Cristo undifferentiated: [11]51–67 Alamitos Formation: [8]63–759 Porvenir Formation: [8]63–246	Madera undifferentiated: [6]11 Alamitos Formation: [8]2.5–30.4 Porvenir Formation: [8]4.2–31.6	--
Suitable	[8]45–70	[8]2.1–2.3	--
--	--	--	--
Suitable (if water found less than 10 ft below land surface)	4 (Unnamed granite-bedrock)[7] 74–1,685 (alluvium and weathered bedrock, undifferentiated)[14]	--	--

[13]Drakos, P , 1997

[14]Shomaker, J W , 1975a

[15]Glorieta Geoscience, Inc , 1996

[16]Lucas, S G , and Hayden, S N , 1991

[17]Shomaker, J W , 1976

[18]Trauger, F D , 1972

[19]Daniel B Stephens & Associates, Inc , 2005

[20]Glorieta Geoscience, Inc , 1986

[21]Durham, L S , 2011

[22]Albright, J L , 1962

[23]Lucas, S G , and Hunt, A P , 1987

[24]Kilmer, L C , 1987

[25]Darr, M , 2011, oral commun

General Structural Geology

During the formation of the present Rocky Mountains (55–80 million years ago), rocks in northwestern parts of San Miguel County were uplifted; these uplifted rocks now form the southern and eastern flanks of the Sangre de Cristo Mountains (Griggs and Hendrickson, 1951). In southwestern parts of the county, uplifting formed "a broad, nearly flat topped arch" now known as Glorieta Mesa (Griggs and Hendrickson, 1951; Baltz, 1972). With increased uplift in western parts of the county, surficial rocks were eroded, and older rocks were exposed at land surface. In central and southeastern parts of San Miguel County, distant from the western uplifted regions, the rock strata remained comparatively undisturbed and generally horizontal (Griggs and Hendrickson, 1951; Baltz, 1972). Trauger (1972) noted, however, that the rocks display a slight eastward regional dip in this region; that towards the northeast the basement dips steeply to form the Las Vegas Subbasin, an arm of the Raton Basin (Woodward and Snyder, 1976); and that the Mesozoic rock sequences thicken and change stratigraphic nomenclature (Mercer and Lappala, 1972; Bejnar and Lessard, 1976; Kilmer, 1987). The rocks in the uplifted western part of the county meet the generally flat-lying rocks in the eastern part of the county along a generally north-south trending transition zone where the rocks are faulted, folded, and fractured, forming north-south trending subsurface anticlines, synclines, and faults (Baltz, 1972). In the transition zone at land surface, folding is expressed in a series of north-south trending hogback monoclinal ridges (Griggs and Hendrickson, 1951); the hogback ridge just west of Las Vegas is known as El Creston, and elsewhere the ridges are referred to as "the southern hogback monocline ridges" (fig. 1B). Folds, fractures, and faults may control the direction and volume of groundwater flow in water-bearing rocks of the transition zone.

Current (2011) Understanding of Hydrogeologic Framework and Groundwater Conditions by Physiographic Area

Groundwater conditions in San Miguel County are described on the basis of Griggs and Hendrickson (1951), updated with recent (March 1972 through November 2010) total well-depth data from the NMWRRS database and aquifer of completion[2] data from the USGS NWIS database (summarized in Matherne and Stewart, 2012), and, where applicable, local-scale data from subdivision and other hydrologic reports archived at the NMOSE Library. Aquifer properties reported in site-specific hydrologic reports are summarized in table 5. Information derived from local-scale hydrologic reports, archived at the NMOSE Library, has been neither verified nor validated.

Total well-depth data are assumed to represent likely water-well drilling depths. Reported total well-depth data are categorized as shallow if they are less than or equal to 300 ft below land surface (bls), moderately deep if they are greater than 300 ft to less than or equal to 600 ft bls, deep if they are greater than 600 to less than or equal to 900 ft bls, and very deep if they are greater than 900 ft bls (table 6). Water

---

[2] "Aquifer of completion" refers to the aquifer formation(s) across which the well is screened and from which the well draws water.

Table 6. Summary of shallow, moderately deep, deep, and very deep New Mexico Water Rights Reporting System (NMWRRS) database wells in each of Griggs and Hendrickson's (1951) physiographic areas and subareas, San Miguel County, New Mexico.

[TD, total depth; bls, below land surface]

Area	Sub-area	Number of wells with TD greater than 0.0 feet bls	Range of reported total depths (feet bls)	Shallow, TD less than or equal to 300 feet bls (percent)	Moderately deep, TD greater than 300 feet bls to less than or equal to 600 feet bls (percent)	Deep, TD greater than 600 feet bls to less than or equal to 900 feet bls (percent)	Very deep, TD greater than 900 feet bls (percent)
1	a, b	747	10–950	77.9	19.3	2.5	0.3
	c, d	612	40–922	60.3	36.0	3.6	0.2
	e	69	39–2,134	72.5	20.3	4.4	2.9
2	a	110	53–1,570	36.4	27.3	29.1	7.3
	b	23	60–995	34.8	39.1	17.4	8.7
3		249	35–880	65.9	28.9	5.2	0.0
4	a	112	24–805	56.3	39.3	4.5	0.0
	b	153	13–2,345	81.1	10.5	6.5	2.0
	c	90	25–1,237	63.3	24.4	5.6	6.7
All wells		2,165		67.3	26.4	5.2	1.1

quality, as summarized by Griggs and Hendrickson (1951, p. 60), is defined arbitrarily as "good" if groundwater samples contained less than 500 parts per million (ppm) of dissolved solids, "fair" if samples contained between 500 and 1,000 ppm of dissolved solids, and "poor" if samples contained more than 1,000 ppm of dissolved solids.

Griggs and Hendrickson (1951) summarized their findings in relation to four major physiographic areas within the county, which generally correspond with the NMOSE declared underground water basins (fig. 5). The Upper Pecos underground water basin generally corresponds with Griggs and Hendrickson's (1951) Areas 1 and 2 except that the northeastern part of Area 1 is located in the Canadian River underground water basin. Griggs and Hendrickson's (1951) Areas 3 and 4 are located generally in the Canadian River and Tucumcari underground water basins, except that the western parts of both areas are located in the Upper Pecos underground water basin (fig. 5). Small portions of the headlands of the Northern Rio Grande and Estancia underground water basins are also present in Area 2.

Griggs and Hendrickson Area 1

The northwestern part of the county includes the southeastern flank of the Sangre de Cristo Mountains. Griggs and Hendrickson (1951) designated this as Area 1 and included subareas 1a–1e (fig. 5) based on the characteristics of water-bearing rocks and associated groundwater availability of each subarea.

In Area 1a, Precambrian crystalline rocks are exposed at land surface, whereas in Area 1b, Sandia Formation and Madera Group sedimentary rocks overlie Precambrian crystalline rocks (Griggs and Hendrickson, 1951). In rocks and alluvium of subareas 1a and 1b, groundwater of good quality in small to moderate quantities occurs at shallow depths (Griggs and Hendrickson, 1951). NMWRRS database well-completion depths in subareas 1a and 1b support the presence of groundwater at shallow depths (table 6). The USGS NWIS database indicates that wells in subarea 1b are completed in the Permian San Andres Formation and Glorieta Sandstone. Wells completed between 1973 and 2010 in subareas 1a and 1b ranged in depth from 10 ft to 950 ft but were generally completed either at shallow depths (about 78 percent) or moderate depths (about 19 percent) and only a few at deep or greater depths (table 6). A review of well-completion data (Matherne and Stewart, 2012) indicates that wells in subareas 1a and 1b generally continue to be completed at shallow depths, as predicted by Griggs and Hendrickson (1951), although between 1973 and 2010 a few wells were completed at depths deeper than predicted. Aquifer characterizations described in subdivision hydrologic reports were consistent with Griggs and Hendrickson's (1951) characterizations of depth and available quantity of groundwater in subareas 1a and 1b (Spiegel, Zane, 1956; Gatlin, 1959; Benjar, 1973, 1974; Shomaker, 1975a, b; New Mexico Soil Conservation Service,

1994; Garn and Jacobi, 1996; Heaton, 1999; MJDarrconsult, Inc., 2003).

Subareas 1c and 1d in the southern part of Area 1 (fig. 5) contain surficial rocks of the Pennsylvanian Sangre de Cristo Formation or of Permian Yeso and San Andres Formations. Reported well depths were generally shallow to moderate, with the Madera Group rocks ranging from near surface to very deep. Small to moderate quantities of fair to good quality groundwater in subareas 1c and 1d may be found either in surficial aquifers (in the Madera Group or Sandia Formation) or in valley alluvium, at shallow to moderate depths (Griggs and Hendrickson, 1951). Aquifers in which wells were reported as completed (Matherne and Stewart, 2012), as well as aquifers identified in local-scale reports (Conover and Murray,1939; Murray, 1944; Sorensen and Gonzales, 1959; New Mexico State Engineer Office, 1960; Benjar, 1974; Shomaker, 1975b; Tolisano and others, 1993; Drakos, 1997; Energia Total, Ltd., 1997; Glorieta Geoscience, Inc., 1997b; Heaton, 1998; Souder, Miller & Associates, 1999; Corbin Consulting, Inc., 2006), are generally consistent with Griggs and Hendrickson (1951) aquifer designations. Wells completed in these subareas between 1972 and 2010 ranged in depth from 40 ft to 922 ft. About 60 percent of the wells were completed at depths of 300 ft or less, about 36 percent were reported as having been completed at moderate depths, while about 4 percent were completed at deep and one well was completed at very deep depths (table 6); these findings are generally consistent with well depths as reported by Griggs and Hendrickson (1951). In subarea 1c, groundwater levels in monitoring well 352725105160901 declined about 12.5 ft between 2006 and 2010 to about 42 ft bls (fig. 8D; location in fig. 5). Groundwater levels in USGS monitoring well 353001105402601, in subarea 1d, showed some fluctuation but averaged about 64 ft bls from 2006 to 2009 (fig. 8C; location in fig. 5). At hogback monoclines, rock strata of all ages may be present and dip steeply toward the east, in places exposing nearly vertical rock outcrops that represent "a wide variety of ground-water conditions" (Griggs and Hendrickson, 1951, p. 61). In subarea 1e, within the rocks of the northern and central hogback monoclines and in alluvial deposits of streams that cut across the ridges, wells produce water at shallow depths in volumes sufficient for domestic use and livestock watering (Griggs and Hendrickson, 1951). About 72 percent of wells completed in subarea 1e between 1974 and 2010 were 300 ft or less in depth, while about 20 percent were moderately deep, about 4 percent were deep, and two wells were very deep (table 6), indicating that most wells produced water from shallow depths as described by Griggs and Hendrickson (1951) but that additional deeper aquifers are at least locally present. One exploration well in subarea 1e, in the area of El Creston west of Las Vegas, is recorded as having been completed at 2,134 ft bls; while it is not known if this well still exists, the record is mentioned to illustrate water-well exploration depths in this subarea.

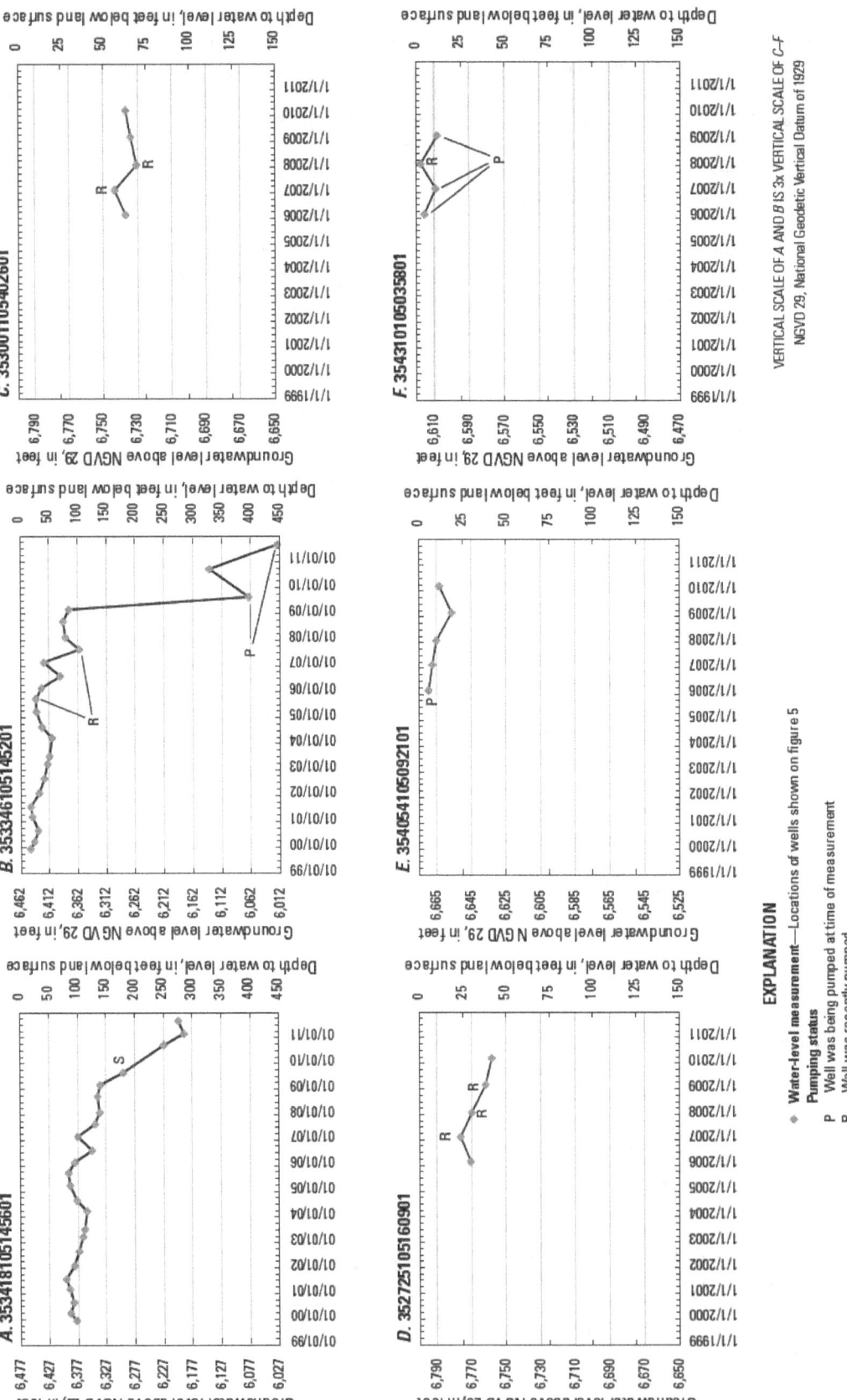

Figure 8.   Groundwater hydrographs for the period of record for U.S. Geological Survey monitoring wells with a period of record of 2 or more years within the vicinity of the Taylor well field, San Miguel County, New Mexico. *A,* 353418105145601. *B,* 353346105145201. *C,* 353001105402601. *D,* 352725105160901. *E,* 354054105092101. *F,* 354310105035801.

Griggs and Hendrickson Area 2

Area 2 is dominated in the southwestern region by Glorieta Mesa, a highland plateau (Griggs and Hendrickson, 1951). The principal water-bearing rocks of Area 2 are the Pennsylvanian Sangre de Cristo Formation, the Permian Yeso Formation, and the Glorieta Sandstone member of the San Andres Formation. According to Griggs and Hendrickson (1951), groundwater in subarea 2a ranges from shallow to very deep and in subarea 2b is generally shallow. The few water-availability reports in Area 2 focused on aquifers contained in Permian and Pennsylvanian rocks (Drakos, 1997; Heaton, 1998; Glorieta Geoscience, Inc., 2003; Corbin Consulting, Inc., 2004); hydrogeologic characteristics derived from these reports are summarized in table 5. Along the western, southern, and eastern flanks of Glorieta Mesa, the Triassic Santa Rosa Sandstone or Chinle Formation rocks may also contain groundwater (table 5). In subarea 2a, wells completed between 1964 and 2010 ranged in depth from 53 to 1,570 feet bls (Matherne and Stewart, 2012). Shallow wells represent a large minority (about 36 percent) of well completions, while deeper well completions were common in subarea 2a, with moderately deep wells accounting for about 27 percent of the total reported completions, deep wells accounting for about 29 percent of well completions, and very deep wells accounting for about 7 percent of well completions (table 6). Wells completed in subarea 2b between 1964 and 2010 ranged from 60 to 995 ft bls (table 6). Well-completion depths in this subarea can be categorized as about 35 percent shallow, 39 percent moderate, 17 percent deep, and 9 percent very deep. Well-completion depths by depth category are generally similar in subareas 2a and 2b for shallow, moderately deep, and deep well completions, indicating that Griggs and Hendrickson's (1951) aquifer depth distinction between subareas of Area 2 may no longer apply. Water quality in Area 2 is generally fair except for water drawn from wells completed in the Yeso Formation, which may contain sulfate at concentrations sufficient to make it unsuitable for domestic use (Griggs and Hendrickson, 1951).

Griggs and Hendrickson Area 3

The western boundary of Area 3 incorporates the southern hogback, which extends between Area 2 and subarea 1e (fig. 5). This part of Area 3 contains the Taylor well field, a supplemental groundwater source to the Las Vegas municipal surface-water supply. Area 3 is generally bounded to the south by the county line and to the north and east by the Canadian Escarpment, the steep ridge that separates the Las Vegas Plateau from the eastern plains (Griggs and Hendrickson, 1951) (fig. 1B), where the land surface has been lowered by erosion due, in part, to flow in tributaries to the Conchas and Canadian Rivers. Conchas Lake, in the east-central part of Area 3, stores waters of these river systems. Area 3 wells were generally completed at depths of 100–300 ft bls and drew fair to good quality water from aquifers of the San Rafael Group and poor to fair quality water from the

Triassic Santa Rosa Sandstone and Chinle Formation (Griggs and Hendrickson 1951). Wells in Area 3 are, however, also completed in alluvium, in terrace or pediment sediments, or in rocks ranging from Pennsylvanian to Triassic age (table 5) (Matherne and Stewart, 2012). Localized hydrologic reports describe groundwater in Area 3 southern hogback monoclinal ridges as found primarily in Permian and Triassic rocks (Albright, 1962; Horner, 1980; Lazarus and Drakos, 1998; Heaton, 2000; John Shomaker & Associates, 2007). East of the hogback monoclinal ridges in Area 3, the nine wells with aquifer completion data in the USGS NWIS database are generally completed in terrace, pediment, or other deposits of gravel, sand, and caliche or in the Triassic Chinle Formation (Matherne and Stewart, 2012). Well completions in the Glorieta Sandstone, Santa Rosa Sandstone, and the Chinle Formation are also reported (Dice, 1954; Schlaikjar, 1971; Trauger, 1972; Shomaker, 1976; U.S. Bureau of Reclamation, 1979).

Between 1972 and 2010, NMWRRS well-completion depths in Area 3 ranged from 35 to 880 ft bls (table 6). About 66 percent of wells were completed at shallow depths, consistent with well depths reported by Griggs and Hendrickson (1951), but about 29 percent of wells were moderately deep and about 5 percent of wells were deep, indicating that groundwater may have been sought at greater depths after 1951.

Near Conchas Lake State Park, groundwater of poor quality but adequate quantity for domestic use was encountered in a well completed at a total depth of about 1,800 ft bls. The well, completed across both the Glorieta and Santa Rosa Sandstones (Shomaker, 1976), was not listed on the NMWRRS dataset downloaded for this report (2010).

Area 3 – Taylor Well Field Area

About 76 percent of wells completed between 1972 and 2010 in Area 3 were completed in the southern hogback area (Matherne and Stewart, 2012), indicating the current focus on groundwater resource development in this area. The Taylor well field, located west of El Creston in the southern hogback area, provides groundwater to Las Vegas (Albright, 1962; Glorieta Geoscience, Inc., 1996; Lazarus and Drakos, 1998). Many detailed hydrologic studies of the area have been performed to characterize hydrologic resources (Heaton, 1935; Jansen, 1935; Northrup and others, 1946; Spiegel, S.J., 1956; Albright, 1962; Molzen–Corbin & Associates, 1980; Molzen–Corbin & Associates and Lee Wilson and Associates, 1985; Glorieta Geoscience, Inc., 1996; Lazarus and Drakos, 1998; Brinkman, 2004; New Mexico Environment Department, 2009, 2010). As early as 1935, Heaton (1935) investigated "groundwater possibilities" in the vicinity of Las Vegas and recommended that an exploratory boring be undertaken at the lowest part of the Las Vegas structural basin, east of El Creston within Las Vegas. Correspondence from Heaton in 1936, included in the NMOSE Library holdings with the 1935 report, indicates that the exploratory well construction was undertaken at the recommended location but that water in quantities sufficient for municipal use was not encountered (Heaton, 1935). S.J. Spiegel (1956) stated that aquifers east of El Creston hogback ridge were likely recharge limited because of structural-geological characteristics of the ridge. S.J. Spiegel (1956) recommended that a test well be constructed southwest of Las Vegas on what is now the Taylor well field, noting that "the strike valleys of low dip provide an area of recharge that is favorable for the accumulation of groundwater, in strata of older age than the Dakota-Entrada horizon." Following S.J. Spiegel's (1956) recommendations, the Taylor well field was established during 1956–57 (Albright, 1962). Albright (1962, p. 23) reported that groundwater supplies from the Taylor well field were sufficient to meet projected needs for Las Vegas through 1990 and recommended acquisition of the Taylor Ranch property to "give protection from close drilling and provide additional locations." A detailed study of the Taylor well field by Glorieta Geoscience, Inc. (1986), which included aquifer and well testing, recommended ongoing maintenance of well-field wells and exploration to locate and prove additional water supplies for Las Vegas. Efforts by Las Vegas to improve well efficiency in the Taylor well field and to identify, prove, and develop new municipal sources of groundwater are ongoing (John Shomaker & Associates, 2007), as are similar efforts among neighboring water users of El Creston and Area 2 (Souder, Miller & Associates, 2010). Groundwater levels have declined markedly in the Taylor well field portion of Area 3 since March 2009, as shown by hydrographs of groundwater elevations for wells completed in the Santa Rosa Sandstone (fig. 8A and B; well locations shown in fig. 5). Groundwater levels in monitoring well 353418105145601 (fig. 8A) declined about 136 ft to a depth of about 276 ft bls between March

2009 and September 2011. Groundwater levels in monitoring well 353346105145201 (fig. 8B) declined about 361 ft to a depth of about 446 ft bls over the same time period, under pumping conditions, as noted in field notes. Groundwater hydrographs of monitoring wells located farther south in the southern hogback area (fig. 9A and B; well locations shown in fig. 5) also display a decline in groundwater levels, with water levels in monitoring well 352949105144301 declining about 38 ft to a depth of about 131 ft bls and in monitoring well 353146105144801 declining about 6.6 ft to a depth of about 50.5 ft bls between September 2010 and October 2011.

Griggs and Hendrickson Area 4

The Las Vegas Plateau rises above the eastern plains along the Canadian Escarpment (Griggs and Hendrickson, 1951) (fig. 1B). The plateau was eroded by action of the Canadian, Conchas, and Pecos Rivers (Griggs and Hendrickson, 1951, p. 14), leaving remnants of the plateau intact; Griggs and Hendrickson referred to these remnants as "outliers" of the main part of the plateau (fig. 1B, fig. 5) and grouped them with the main plateau into Area 4: "Las Vegas Plateau and outliers" (Griggs and Hendrickson, 1951). Area 4 is divided into three subareas. Subarea 4a, the main portions and outlying remnants of the Las Vegas Plateau, are found in the eastern two-thirds of the county (fig. 5), while subareas 4b and 4c are located in the north-central part of the county. Although Las Vegas (fig. 1A) is located in the southwestern part of subarea 4b, Las Vegas' Taylor well field is located in Area 3 (fig. 5). Subarea 4a wells were generally completed in the early-Cretaceous Dakota Sandstone at depths of 250 ft bls or less. In subarea 4b, wells were completed in the mid-Cretaceous Graneros and Greenhorn Limestone Members of the Mancos Shale at depths of 250 ft bls or less.

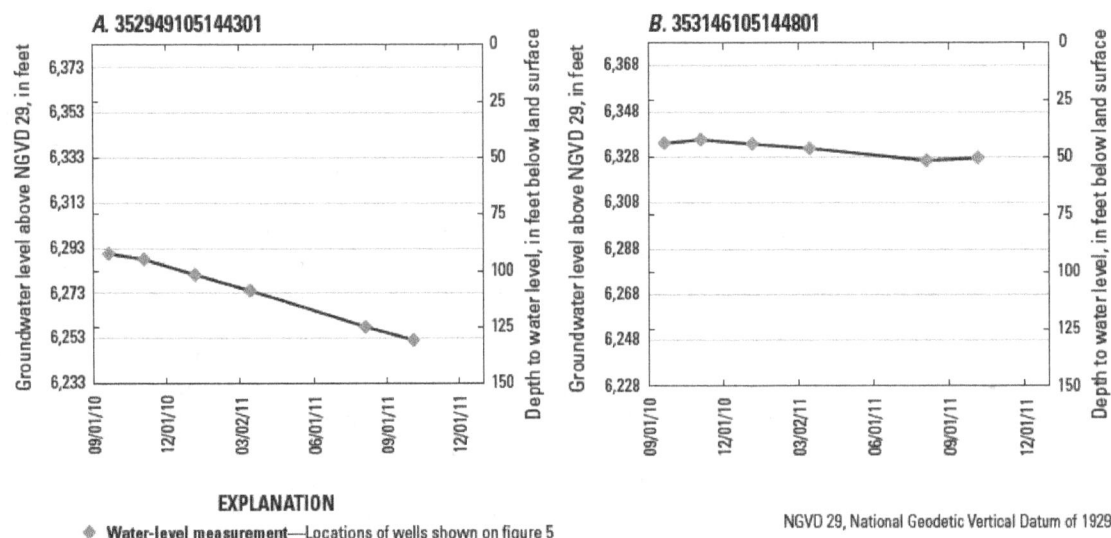

EXPLANATION

◆ Water-level measurement—Locations of wells shown on figure 5

NGVD 29, National Geodetic Vertical Datum of 1929

Figure 9. Groundwater hydrographs for U.S. Geological Survey monitoring wells in the southern hogback area, San Miguel County, New Mexico. A, 352949105144301. B, 353146105144801.

In subarea 4c, wells were completed in the late Cretaceous Carlile and Niobrara Members of the Mancos Shale (Griggs and Hendrickson, 1951) (table 5). Some reviewed studies recommended drilling into deeper, older rocks if groundwater of sufficient quantity was not found in shallow rocks (Heaton, 1935; Griggs and Hendrickson, 1951; Spiegel, S.J., 1956).

Between 1973 and 2010, approximately 56 percent, 81 percent, and 63 percent of wells were completed at depths of 300 ft bls or less in subareas 4a, 4b, and 4c, respectively (table 6), consistent with well depths reported by Griggs and Hendrickson (1951). Moderately deep well completions were more prevalent in subarea 4a than in subareas 4b or 4c (about 39 percent, 11 percent, and 24 percent, respectively), whereas few deep and very deep well completions were reported (in combination, about 5 percent, 8.5 percent, and 12 percent, respectively) (table 6). Water levels in USGS monitoring wells in subarea 4b (fig. 8E and F; locations shown in fig. 5) show some fluctuation but averaged about 11 ft bls for the period 2006–10 for well 354054105092101 (fig. 8E) and about 7 ft bls for the period 2006–9 for well 354310105035801 (fig. 8F).

Summary of Hydrogeologic Framework and Groundwater Conditions by Physiographic Area

Data compiled since Griggs and Hendrickson's (1951) report on the geology and groundwater resources of San Miguel County are generally consistent with Griggs and Hendrickson's (1951) original characterization of depth and availability of groundwater resources and of source aquifers. Subsequent exploratory drilling identified deep available groundwater in Area 3 (Shomaker, 1976) but did not identify groundwater sufficient for municipal use in Area 4c, in the vicinity of Las Vegas (Matherne and Stewart, 2012). Most current (2011) development of groundwater resources is in western San Miguel County, particularly in the vicinity of El Creston, where USGS groundwater-monitoring wells indicate that groundwater levels are declining (fig. 8; locations shown in fig. 5). Most wells in the eastern part of the county are limited to domestic and stock wells, and little information is available regarding additional groundwater resources in that area.

Data Gaps and Suggestions for Future Study

Publicly available hydrologic reports and datasets were used to summarize the current understanding of the hydrologic resources of San Miguel County and to identify gaps in hydrologic data and understanding necessary for informed management of county-wide water resources. The Region 8 Water Plan, which considered San Miguel, Mora, and Guadalupe Counties, was one of the reports reviewed for the present work (Daniel B. Stephens & Associates, Inc., 2005). The Region 8 Water Plan represents findings of a 5-year

assessment process, which included a series of community meetings to obtain public input, that characterized the water supply, projected future demand, and presented alternatives for meeting projected demand throughout the three-county study area (Daniel B. Stephens & Associates, Inc., 2005). The present summary of available literature and data for San Miguel County supports the findings of the Region 8 Water Plan with respect to data gaps identified in San Miguel County and also with respect to potential projects that could assist San Miguel County managers in assessing and planning for future needs (Daniel B. Stephens & Associates, Inc., 2005).

Data Gaps Identified and Suggestions for Future Study To Assess Surface-Water Resources

The ability to evaluate and quantify surface-water resources, both as runoff and as potential groundwater recharge, could be enhanced by expanding the network of measurement stations throughout the county. Currently, high-elevation precipitation is monitored by a single SNOTEL station at Wesner Springs (fig. 2). Additional SNOTEL stations in the Sangre de Cristo Mountains and Glorieta Mesa could improve spatial and temporal estimates of high-elevation precipitation and snow pack and could be used to improve estimates of potential recharge and runoff for irrigation, reservoir storage, and management.

At lower elevations, although 19 USGS streamgages have been operated at various times within San Miguel County, only 5 continuous streamgages are currently in use (fig. 2), so that the ability to monitor and quantify contributions from tributaries and within the major river basins is limited. Tecolote Creek, a major tributary of the upper Pecos River, is currently gaged for peak flows only (fig. 2). The Conchas River above Conchas Lake and the Canadian River below the reservoir are not gaged within San Miguel County. Streamgages on these stream reaches within San Miguel County, and more detailed monitoring of the timing and volume of surface-water withdrawals and return flows for municipal and irrigation purposes, could contribute to quantification of volumes and sources of available surface water within the county and the quantification of channel transmission losses and associated focused groundwater recharge.

Data Gaps Identified and Suggestions for Future Study To Assess Surface-Water/Groundwater Interactions

Flow characteristics of the Pecos, Gallinas, and Canadian Rivers with respect to locations and potential quantities of recharge to groundwater, or contributions of groundwater to base flow, have not been systematically defined, based on the review of available literature. In particular, the magnitude of recharge to groundwater from unlined diversion canals and return to base flow from irrigation and municipal use, the potential effects of groundwater withdrawals as flow loss captured from nearby streams, and contributions of mountain-front groundwater recharge are components of the groundwater/surface-water system that could be improved with further delineation and quantification.

Only a few seepage surveys have been performed in San Miguel County (Jansen, 1935; U.S. Geological Survey, 1978, 2006). A series of additional seepage surveys along the lengths of the rivers could help to determine locations of surface-water losses to and gains from the local groundwater system and identify reaches of interest for more detailed study. Seepage surveys coinciding with irrigation and nonirrigation seasons could help to quantify the component of streamflow attributable to irrigation return flow; associated synoptic water-quality sampling could help to identify potential effects to water quality attributable to irrigation return flow. Specific seepage surveys along diversion canals could help quantify groundwater recharge from unlined diversion canals.

Studies characterizing the effects of groundwater withdrawals on streamflow in San Miguel County were not identified in the assessment of available literature on which this report is based. To assess such effects, monitoring wells could be constructed along transects between production wells and stream reaches of interest to monitor decline or recovery of the water table, to quantify the timing and extent of water-table response, and to identify the spatial extent of capture zones. Streamflow discharge measurements upstream and downstream from capture zones could help to define flow loss or gain in the vicinity of production wells to aid in assessment of gains or losses along the reach. A series of such studies could provide necessary information for conjunctive management of surface-water and groundwater supplies in water-critical areas of the county.

The spatial distribution and quantity of mountain-front and mountain-block groundwater recharge are currently unknown; these data could be estimated theoretically by using precipitation data, land use coverages, topography, digital elevation models, and hydrogeologic properties estimated from geologic maps and available aquifer tests. Systematic acquisition of such recharge data would allow local water-balance estimates to be expanded to a county-wide scale.

Data Gaps Identified and Suggestions for Future Study To Assess Groundwater Resources

The literature review for this study indicated that there is a county-wide lack of systematic information about depth to water and associated potential for groundwater development. While it is reasonable to assume that shallow groundwater flow in San Miguel County, as elsewhere, generally follows topography, existing data are not sufficient to construct groundwater-flow maps for the county. Aside from the Griggs and Hendrickson (1951) county-wide study, groundwater investigations have been localized and tied to suburban

development projects, although some investigations have located sources of groundwater at depths deeper than early drilling efforts in some areas of the county. Hydrogeologic investigations and well drilling have been generally focused on the western third of the county, where population density is greatest, and information about groundwater resources in the eastern two-thirds of the county is sparse. Assessment of groundwater occurrence and location could be aided by a county-wide distribution of water-level information and by a series of maps of groundwater potential, compiled for each individual aquifer, including saline aquifers, for which the potential for municipal use through desalination could be explored.

Two steps necessary to address the identified data needs have been met by the present USGS effort. First, in cooperation with the NMOSE, the groundwater-monitoring network has been expanded, by using existing domestic or livestock wells, to include representation across San Miguel County (fig. 5). The expanded groundwater-monitoring network includes 33 wells, most measured on a 5-year schedule. Annual or biannual monitoring could increase the utility of the monitoring network in documenting and tracking changes in groundwater levels throughout the county. Second, results of local hydrogeologic studies, including reported aquifer properties, have been compiled in this report to update understanding of the hydrogeologic framework of San Miguel County (table 5). The expanded groundwater-monitoring network and the updated hydrogeologic framework form the first two steps in developing regional-scale water-level (potentiometric) maps for individual aquifers in San Miguel County.

Well logs from well drillers' reports provide spatially related hydrologic data, including total well depth, first encounter with groundwater, water level after well completion, strata encountered, and results of aquifer testing, if any tests were performed. The aquifer of completion is generally not described by the well driller; however, the aquifers of completion were interpreted and correlated for a small number of the NMWRRS database wells in the USGS NWIS database for wells drilled between 1991 and 1996 (Matherne and Stewart, 2012) . This correlation effort could be renewed and expanded in selected areas of interest for groundwater development by using information from the drillers' reports to identify the aquifer of completion and to estimate aquifer productivity and hydraulic characteristics of the aquifers of interest. A limited number of drillers' reports would have sufficient information to make these determinations. Providing that a sufficient number of wells were found to be completed in given aquifers of interest, potentiometric maps for these aquifers could be developed. Well location information for many wells used in the present study was approximated by well drillers using techniques available at the time of drilling. Estimation and mislocation of well locations introduce an unknown amount of uncertainty in interpretive results, and verification and resurvey of well locations by using Global Positioning System (GPS) technology should be part of an effort to develop potentiometric maps based on information derived from drillers' reports. Water levels could be measured when wells are resurveyed to develop current potentiometric maps for individual aquifers. Once developed, potentiometric maps could be used to characterize regional-scale aquifers and would provide information that could be used to generate a conjunctive (combined surface-water/groundwater) county-wide water balance, as well as to help identify new areas and depths for test drilling to further delineate the depth of groundwater for unique aquifers.

Change in the volume of groundwater storage in the vicinity of the Taylor well field (fig. 8) has increased with increased development and consequent increased reliance on local groundwater resources. Microgravity measurements have proven useful in tracking temporal changes in volumetric groundwater storage in the semiarid southwestern United States (Pool and Schmidt, 1997; Parker and Pool, 1998); small changes in the strength of the local gravitational field are measured at gridded locations along a predefined network at time intervals of interest (for example, annually). Temporal differences observed in the local gravitational field are attributed to changes in the volume, or total mass, of groundwater stored over time. Microgravity techniques can quantify changes in groundwater storage but cannot quantify the absolute volume of stored groundwater. Nevertheless, microgravity studies may be useful in San Miguel County where quantification of the change in groundwater storage in areas such as the Taylor well field vicinity is of sufficient importance to justify the acquisition of specialized equipment, site setup, and long-term microgravitational-field monitoring.

Data Gap Identified and Suggestions for Future Study: A Comprehensive Geographic Information System Hydrologic Geodatabase

Development of a county-wide geographic information system (GIS) hydrologic geodatabase could provide a decision-support tool for future water-management decisions. Such a geodatabase could provide a place to aggregate, store, and manage hydrologic data and could provide a means to analyze, map, and display data spatially and temporally. For example, conjunctive water balances could be computed at scales from subbasin to municipal levels. The ability to perform hydrologic geospatial analyses would be tied to the scale and availability of the underlying data. Such a tool could aid in archiving and updating data, with associated uncertainty, both spatially and temporally; could allow ongoing assessment of spatial and temporal data gaps; and could allow information-gathering efforts to be focused to support water-management decisions.

Hydrologic data not currently in digital format could be converted to digital format and stored in the hydrologic geodatabase with associated location information and metadata. Data layers that should be included in a geodatabase include surficial geology, elevation and topographic

information, soil types, land use and land cover, the hydrologic network of rivers and lakes, irrigation canals and diversions, streamgages and crest-stage gages, climate stations, validated NMOSE well locations, the USGS groundwater-monitoring network, spring locations, water-quality data collection points and tabulated results, and a bibliography of records pertaining to San Miguel County associated with respective study areas. Data layers could include hyperlinks to online reports and records, such as water levels and water-quality data, where possible.

Summary

The surface-water and groundwater resources of San Miguel County, New Mexico, are increasingly relied upon to meet growing domestic, livestock, and agricultural needs. San Miguel County is an area with an expanding economy and population, and to meet future water demands, aquifers may be further developed. The U.S. Geological Survey (USGS), in cooperation with San Miguel County, conducted a study to assess publicly available information regarding the hydrologic resources of San Miguel County and to identify data gaps in that information and hydrologic information that could aid in the management of available water resources.

San Miguel County comprises four physiographic areas: the Sangre de Cristo Mountains in the northwest, Glorieta Mesa in the southwest, the Las Vegas Plateau and outliers in the north-central and eastern areas, and the plains and southern hogback (monocline) east of the Sangre de Cristo Mountains and Glorieta Mesa. The types of principal aquifers in the study area vary by physiographic area. San Miguel County is drained primarily by the Pecos and Canadian Rivers. Seventy-four percent of the land of San Miguel County, about 3,500 mi², is agricultural, of which about 92 percent is in pasture. The county seat, Las Vegas, had a 2010 population of about 13,750, or about 20,000 when the surrounding area was included. Projected population increase is primarily along the Interstate 25 corridor between Santa Fe and Las Vegas. Surface water supplies about 97 percent of the water demand in San Miguel County.

Mean monthly and annual discharge at USGS streamgages on the Pecos River generally increases downstream. Mean monthly discharge is generally bimodally distributed, with about 47 percent of annual streamflow occurring during April through June, corresponding to spring runoff, and with a lesser peak in August, representing about 11 percent of annual streamflow, corresponding to summer monsoonal rains. Flow duration curves based on mean daily discharge values for the period 1981–2010 indicate that mean daily discharge for the middle 90 percent of discharge values ranges over two to three orders of magnitude, reflecting higher flows during spring and summer runoff as compared to flows during the winter months. Mean monthly discharge on the Gallinas Creek near Montezuma is bimodally distributed, with 38 percent of the mean annual discharge in April and May, corresponding to spring runoff, and a lesser peak in August representing 14 percent of the mean annual discharge and corresponding to summer monsoonal rains. Mean monthly discharge on the Gallinas River near Colonias shows a subdued response to spring runoff, with 47 percent of runoff in July and August corresponding to summer monsoonal rain. The low-flow area of the Gallinas Creek near Montezuma curve is relatively flat compared to the Gallinas River near Colonias curve, indicating more sustained base flow or snowmelt runoff than is indicated by the somewhat steeper slope of the Gallinas River near Colonias curve. Flow in Gallinas Creek near Montezuma is perennial and has not been measured at less than 1.2 cubic feet per second during the period of record. The Gallinas River near Colonias is characterized by periods of declining base flow or snowmelt runoff, with periodic high-flow events and periods of no flow. Mean monthly discharge at the Canadian River near Sanchez is bimodally distributed, with about 29 percent of flow in May through June, corresponding to spring runoff, and 15 percent in August, corresponding to summer monsoonal rains. The gradual decline of the flow duration curve for the Canadian River near Sanchez in the middle 90 percent of the curve likely reflects greater temporal and spatial variability with respect to precipitation, runoff, and potential transmission losses associated with the larger and more diverse contributing source area, as compared to the contributing source areas for the Pecos and Gallinas Rivers and tributaries.

Within San Miguel County, approximately 270 river miles of the Pecos Headwaters watershed and 100 river miles of the upper Canadian watershed were listed as impaired by the New Mexico Environment Department. Impairments were primarily for the uses High Quality Cold Water Aquatic Life and Marginal Cold Water Fisheries. Criteria for impairment were primarily temperature and turbidity but also included nutrients, ammonia, nitrate and nitrite, pH, dissolved oxygen, specific conductance, siltation, and mercury in fish tissue in selected reaches.

Data compiled since 1951 on the geology and groundwater resources of San Miguel County are generally consistent with the original characterization of depth and availability of groundwater resources and of source aquifers. Subsequent exploratory drilling identified deep available groundwater in some locations. Most current (2011) development of groundwater resources is in western San Miguel County, particularly in the vicinity of El Creston, where USGS groundwater-monitoring wells indicate that groundwater levels are declining. Most wells in the eastern part of the county are limited to domestic and stock wells, and little information is available regarding additional groundwater resources in that area.

Regarding future studies to address identified data gaps, the ability to evaluate and quantify surface-water resources, both as runoff and as potential groundwater recharge, could be enhanced by expanding the network of measurement stations throughout the county. Additional SNOTEL stations in the Sangre de Cristo Mountains and Glorieta Mesa could improve spatial and temporal estimates of high-elevation precipitation and snow pack and could be used to improve estimates of potential recharge and runoff for irrigation, reservoir storage, and management. At lower elevations, additional streamgages within San Miguel County, and more detailed monitoring of the timing and volume of surface-water withdrawals and return flows, could contribute to quantification of volumes and sources of available surface water within the county and the quantification of channel transmission losses and associated focused groundwater recharge.

A series of seepage surveys along the lengths of the rivers could help to determine locations of surface-water losses to and gains from the local groundwater system and identify reaches of interest for more detailed study. Seepage surveys coinciding with irrigation and nonirrigation seasons could help to quantify the component of streamflow attributable to irrigation return flow; associated synoptic water-quality sampling could help to identify potential effects to water quality attributable to irrigation return flow. Studies characterizing the effects of groundwater withdrawals on streamflow in San Miguel County were not identified in the assessment of available literature on which this report is based. To assess such effects, monitoring wells could be constructed along transects between production wells and stream reaches of interest to monitor decline or recovery of the water table, to quantify the timing and extent of water-table response, and to identify the spatial extent of capture zones.

Hydrogeologic investigations and well drilling have been generally focused on the western third of the county, where population density is greatest, and information about groundwater resources in the eastern two-thirds of the county is sparse. Assessment of groundwater potential could be aided by a county-wide distribution of water-level information and by a series of maps of groundwater potential, compiled for each individual aquifer, including saline aquifers, for which the potential for municipal use through desalination could be explored. Once developed, potentiometric maps could be used to characterize regional-scale aquifers and would provide information that could be used to generate a conjunctive (combined surface-water/groundwater) county-wide water balance, as well as to help identify new areas and depths for test drilling to further delineate the depth of groundwater for unique aquifers.

Development of a county-wide geographic information system (GIS) hydrologic geodatabase could provide San Miguel County with a decision-support tool for future water-management decisions. Such a geodatabase could provide a place to aggregate, store, and manage hydrologic data and could provide a means to analyze, map, and display data spatially and temporally.

Bibliography

Documents consulted for this report are from publicly available sources including libraries, the U.S. Geological Survey Publications Warehouse, the New Mexico Office of the State Engineer Library, and common online search engines.

Aguirre, L.P., 2008, Water master report, Gallinas River, 2008: Santa Fe, N. Mex., New Mexico Office of the State Engineer, Interstate Stream Commission, 46 p., accessed November 10, 2011, at http://www.ose.state. nm.us/publications/WaterMasterReports/Gallinas/ Gallinas2008Report.pdf.

Aguirre, L.P., 2009, Water master report, Gallinas River, 2009: Santa Fe, N. Mex., New Mexico Office of the State Engineer, Interstate Stream Commission, 32 p., accessed November 10, 2011, at http://www.ose.state. nm.us/publications/WaterMasterReports/Gallinas/ Gallinas2009Report.pdf.

Aguirre, L.P., 2010, Water master report, Gallinas River, 2010: Santa Fe, N. Mex., New Mexico Office of the State Engineer, Interstate Stream Commission, 36 p., accessed November 10, 2011, at http://www.ose.state. nm.us/publications/WaterMasterReports/Gallinas/ Gallinas2010Report.pdf.

Albright, J.L., 1962, Groundwater supply and geology, Las Vegas, New Mexico: Prepared for the Public Service Company of New Mexico, on file at the Library of the New Mexico Office of the State Engineer, 24 p.

Baltz, E.H., 1972, Geologic map and cross sections of the Gallinas Creek area, Sangre De Cristo Mountains, San Miguel County, New Mexico: U.S. Geological Survey, Miscellaneous Geologic Investigations Map I-673, scale 1:24,000. (Also available at http://ngmdb.usgs.gov/Prodesc/ proddesc_9450.htm.)

Baltz, E.H., and Myers, D.A., 1984, Porvenir Formation (new name) and other revisions of nomenclature of Mississippian, Pennsylvanian, and lower Permian rocks, southeastern Sangre de Cristo Mountains, New Mexico: U.S. Geological Survey Bulletin 1537–B, Contributions to Stratigraphy, 39 p. (Also available at http://pubs.usgs.gov/bul/1537b/ report.pdf.)

Banta, E.H., 1963, Water master report, Pecos Valley Surface Water District: Santa Fe, N. Mex., New Mexico Office of the State Engineer, Interstate Stream Commission, on file at the Library of the New Mexico Office of the State Engineer, 41 p.

Bejnar, C.R., and Lessard, R.H., 1976, Paleocurrents and depositional environments of the Dakota Group, San Miguel and Mora Counties, New Mexico in Ewing, R.C., and Kues, B.S., eds., Vermejo Park, northeastern New Mexico—New Mexico Geological Society, 27th annual field conference, September 30–October 1–2, 1976: New Mexico Geological Society, p. 157–163.

Benjar, Waldemere, 1973, Geologic and ground-water report of El Rancho del Canoncito area, San Miguel County, New Mexico: Prepared for Holiday Properties, Inc., on file at the Library of the New Mexico Office of the State Engineer, 20 p.

Benjar, Waldemere, 1974, Geologic and ground-water report of Hermit's Peak Ranch area, San Miguel County, New Mexico: Prepared for Hermit's Peak Ranches, Inc., on file at the Library of the New Mexico Office of the State Engineer, 35 p.

Berger, A.C., Bethke, C.M., and Krumhansl, J.L., 2000, A process model of natural attenuation in drainage from a historic mining district: Applied Geochemistry, v. 15, p. 655–666.

Brinkman, James, 2004, Evaluation of groundwater use and the effects of pumping on groundwater levels in the El Creston Area: Prepared for the Valle Viejo Water Users Association, on file at the Library of the New Mexico Office of the State Engineer, 36 p., 4 apps.

Chavez, M.A., 2004, Watermaster report, Gallinas River, 2004: Santa Fe, N. Mex., New Mexico Office of the State Engineer, Interstate Stream Commission, 14 p., accessed November 10, 2011, at http://www.ose.state. nm.us/publications/WaterMasterReports/Gallinas/ Gallinas2004Report.pdf.

Committee on the Scientific Basis of Colorado River Basin Water Management, 2007, Colorado River Basin water management—Evaluating and adjusting to hydroclimatic variability: Washington, D.C., The National Academies Press, 210 p. (Also available at http://www nap.edu/catalog. php?record_id=11857.)

Conover, C.S., and Murray, C.R., 1939, Report on the artesian well at San Miguel, San Miguel County, New Mexico: On file at the Library of the New Mexico Office of the State Engineer, 4 p.

Corbin Consulting, Inc., 2004, Geohydrology report—"Tierra de Dios" Subdivision, (section 1, T12N, R12E), San Miguel County, New Mexico: On file at the Library of the New Mexico Office of the State Engineer, 15 p.

Corbin Consulting, Inc., 2006, Geohydrology report— George Roybal Property, (section 31, T16N, R12E), San Miguel County, New Mexico: On file at the Library of the New Mexico Office of the State Engineer, 16 p., 7 apps.

Cunningham, W.L., and Schalk, C.W., comps., 2011, Groundwater technical procedures of the U.S. Geological Survey: U.S. Geological Survey Techniques and Methods 1–A1, 151 p., accessed May 29, 2011, at http://pubs.usgs. gov/tm/1a1/.

Daniel B. Stephens & Associates, Inc., 2005, Mora-San Miguel-Guadalupe regional water plan: Prepared for New Mexico Office of the State Engineer, Interstate Stream Commission, 377 p., accessed September 2, 2011, at http://www.ose.state.nm.us/isc_regional_plans8 html.

Davis, D.R., 2000, Water quality and biological assessment survey of Monastery Lake, San Miguel County, July 11, 2000 in Surface Water Quality Bureau, eds., Water quality assessments for selected New Mexico Lakes, 2000: Santa Fe, N. Mex., Monitoring and Assessment Section, Surface Water Quality Bureau, New Mexico Environment Department, 20 p., accessed August 18, 2011, at http://www.nmenv.state.nm.us/swqb/Surveys/LakeWaterQualityAssessments2000.pdf.

Dice, H.E., 1954, Report by watermaster of Gallinas River Water District for combined seasons of 1953 and 1954: Santa Fe, N. Mex., New Mexico Office of the State Engineer, on file at the Library of the New Mexico Office of the State Engineer, 15 p.

Drakos, Paul, 1997, Geohydrology of Vistas del Valle Subdivision, San Miguel County, New Mexico: Santa Fe, N. Mex., Glorieta Geoscience, Inc., on file at the Library of the New Mexico Office of the State Engineer, 23 p., 6 apps.

Dunne, Thomas, and Leopold, L.B., 1978, Water in environmental planning: San Francisco, W.H. Freeman, 818 p.

Durham, L.S., 2011, New plays cropping up—Sprawling Niobrara has multiple models: AAPG Explorer, 2 p., accessed October 2011 at http://www.aapg.org/explorer/2011/06jun/niobrara0611.cfm.

Ebright, Malcolm, 2009, Storrie Lake State Park—History of title and history of the town of Las Vegas land grant: Prepared for the Commission for Public Records, 61 p., accessed October 15, 2011, at http://www.newmexicohistory.org/featured_projects/Legislative%20Reports/documents/StorrieLake.pdf.

Energia Total, Ltd., 1997, Santa Fe Trail Ranch, water supply report: On file at the Library of the New Mexico Office of the State Engineer, 19 p.

Flynn, K.M., Kirby, W.H., and Hummel, P.R., 2006, User's manual for program PeakFQ annual flood frequency analysis using Bulletin 17B guidelines: U.S. Geological Survey Techniques and Methods, book 4, chap. B4, 42 p. (Also available at http://pubs.usgs.gov/tm/2006/tm4b4/.)

Freeze, R.A., and Cherry, J.A., 1979, Groundwater: Englewood Cliffs, N.J., Prentice-Hall, 604 p.

Garn, H.S., and Jacobi, G.Z., 1996, Water quality and benthic macroinvertebrate bioassessment of Gallinas Creek, San Miguel County, New Mexico, 1987-90: U.S. Geological Survey Water Resources Investigations Report 96–4011, 62 p. (Also available at http://pubs.usgs.gov/wri/1996/4011/report.pdf.)

Garner, W.L., 1956, Report of water master of Pecos River Water District for season of 1956: New Mexico Office of the State Engineer, on file at the Library of the New Mexico Office of the State Engineer, 33 p.

Gatlin, J.C., 1959, Gallinas Project, Pecos River Basin, New Mexico: Prepared for the Bureau of Reclamation, Amarillo, Tex., Bureau of Sport Fisheries and Wildlife technical memorandum, on file at the Library of the New Mexico Office of the State Engineer, 3 p.

Glorieta Geoscience, Inc., 1986, Geologic and hydrologic survey of the Taylor well field and vicinity, Las Vegas, San Miguel County, New Mexico: Prepared for the City of Las Vegas and Bohannan-Huston, Inc., on file at the Library of the New Mexico Office of the State Engineer, 64 p., 7 apps.

Glorieta Geoscience, Inc., 1991, Mora San Miguel regional water study and forty year plan, summary report: Prepared for the Mora San Miguel Water Plan Committee, on file at the Library of the New Mexico Office of the State Engineer, 14 p.

Glorieta Geoscience, Inc., 1996, Pumping test results Taylor well field 17-day pumping test and 27-day recovery test: Prepared for the City of Las Vegas, N. Mex., on file at the Library of the New Mexico Office of the State Engineer, 3 p.

Glorieta Geoscience, Inc., 1997a, Geohydrology of Padre Springs Subdivision, San Miguel County, New Mexico: On file at the Library of the New Mexico Office of the State Engineer, 21 p., 5 apps.

Glorieta Geoscience, Inc., 1997b, Return flow credit plan for the Village of Pecos UP-86 et al.: Prepared for the Village of Pecos, on file at the Library of the New Mexico Office of the State Engineer, 14 p., 4 apps.

Glorieta Geoscience, Inc., 2003, Geohydrology of the Luz del Sol (II) Subdivision (5 lot addition to existing Luz del Sol Subdivision), San Miguel County, New Mexico: Prepared for Winstar, Inc., on file at the Library of the New Mexico Office of the State Engineer, 22 p., 4 apps.

Glorieta Geoscience, Inc., and James W. Siebert & Associates, 1990, Hydrologic assessment of the Mora-San Miguel water plan area: Prepared for the Mora-San Miguel Water Plan Committee, on file at the Library of the New Mexico Office of the State Engineer, 49 p.

Green, Gregory, and Jones, Glen, 1997, The digital geologic map of New Mexico in ARC/INFO format, in USGS Geoscience Data Catalog: U.S. Geological Survey Open-File Report 97–52, accessed June 10, 2011, at http://geo-nsdi.er.usgs.gov/metadata/open-file/97-52/metadata faq.html.

Griggs, R.L., and Hendrickson, G.E., 1951, Geology and ground-water resources of San Miguel County, New Mexico: New Mexico Bureau of Mines and Mineral Resources, Groundwater Report 2, 121 p.

Heath, Ralph, 1983, Basic ground-water hydrology: U.S. Geological Survey Water-Supply Paper 2220, 86 p. (Also available at http://pubs.usgs.gov/wsp/2220/report.pdf.)

Heaton, C.M., 1998, Reconnaissance hydrology report Mesa Vista Property, Township 14N Range 12E section 13, San Miguel County, New Mexico: Santa Fe, N. Mex., SinAgua Consultants, on file at the Library of the New Mexico Office of the State Engineer, 20 p.

Heaton, C.M., 1999, Groundwater hydrology report Bonnie McGowan property, township 16 North, Range 12 East, section 25, San Miguel County, New Mexico: Santa Fe, N. Mex., SinAgua Consultants, on file at the Library of the New Mexico Office of the State Engineer, 8 p.

Heaton, C.M., 2000, Aquifer hydrology report—Amadeo Tenorio Jr. property, township 14 North, Range 16 East, section 14, San Miguel County, New Mexico: Santa Fe, N. Mex., SinAgua Consultants, on file at the Library of the New Mexico Office of the State Engineer, 4 p.

Heaton, R.L., 1935, Geology and water possibilities of a portion of the Las Vegas Basin: Prepared for New Mexico Power Company, Las Vegas, N. Mex., on file at the Library of the New Mexico Office of the State Engineer, 9 p.

Hem, J.D., 1952, Quality of water, Conchas Reservoir, New Mexico, 1939–49: U.S. Geological Survey Water Supply Paper 1110-C, 56 p. (Also available at http://pubs.usgs.gov/wsp/1110c/report.pdf.)

Hilley, T.E., McCarty, T.E., Martin, P.G., and Sellnow, S.L., 1981, Soil survey of San Miguel County area, New Mexico: U.S. Department of Agriculture Soil Conservation Service and Forest Service, 169 p. (Also available at http://soils.usda.gov/survey/online_surveys/new_mexico/.)

Hopkins, J.S., 2001, Water quality assessment of the Gallinas River and Tecolote Creek: Santa Fe, N. Mex., Monitoring and Assessment Section, Surface Water Quality Bureau, New Mexico Environment Department, 12 p., accessed September 2, 2011, at http://www.nmenv.state nm.us/swqb/Surveys/UpperPecosPartII-2001.pdf.

Horn, Marty, and Timmons, J.M., 2006, Preliminary geologic map of the Ojitos Frios quadrangle, San Miguel County, New Mexico: New Mexico Bureau of Geology and Mineral Resources, Open-File Map OFGM-130.

Horner, W.P., 1980, Water availability Martinez tact, Las Vegas-Romeroville area, San Miguel County: On file at the Library of the New Mexico Office of the State Engineer, 5 p.

Jansen, E.C., 1935, Investigation of the water supply of Las Vegas, New Mexico: Denver, Colo., Public Service Company of Colorado, on file at the Library of the New Mexico Office of the State Engineer, 20 p.

John Shomaker & Associates, 2007, Summary of information relating to wells, pumping tests and water quality high yield wells on Millikin Ranch, Las Vegas, New Mexico: Prepared for Millikin Ranch, on file at the Library of the New Mexico Office of the State Engineer, 16 p., 3 apps.

Johnson, P.S., and Deeds, J.L., 1995a, Summary of environmental issues at El Molino mill, north-central New Mexico, in Bauer, P.W., Kues, B.S., Dunbar, N.W., Karlstrom, K.E., and Harrison, Bruce, eds., Geology of the Santa Fe region—New Mexico Geological Society, 46th annual field conference, September 27–30, 1995: New Mexico Geological Society, p. 319–322.

Johnson, P.S., and Deeds, J.L., 1995b, A site conceptual model of environmental issues at the Pecos mine, north-central New Mexico, in Bauer, P.W., Kues, B.S., Dunbar, N.W., Karlstrom, K.E., and Harrison, Bruce, eds., Geology of the Santa Fe region—New Mexico Geological Society, 46th annual field conference, September 27–30, 1995: New Mexico Geological Society, p. 41–43.

Kilmer, L.C., 1987, Water-bearing characteristics of geologic formations in northeastern New Mexico – southeastern Colorado, in Lucas, S.G., and Hunt, A.P., eds., Northeastern New Mexico—New Mexico Geological Society, 38th annual field conference, September 24–26, 1987: New Mexico Geological Society, p. 275–279.

King, Norman, 1974, The Carlile-Niobrara contact and lower Niobrara strata near El Vado, New Mexico in Woodward, L.A., and Callender, J.F., eds., Ghost Ranch (central-northern N.M.)—New Mexico Geological Society, 25th annual field conference, October 10–12, 1974: New Mexico Geological Society, p. 259–266. (Also available at http://nmgs nmt.edu/publications/guidebooks/downloads/25/25_p0259_p0266.pdf.)

Lazarus, Jay, and Drakos, P.G., 1998, Geohydrologic characteristics of the Taylor well field, City of Las Vegas, New Mexico, in Annual New Mexico Water Conference, 42d, Tucumcari, N. Mex., 1997, Proceedings: New Mexico Water Resources Research Institute report no. 304, p. 59–71. (Also available at http://wrri nmsu.edu/publish/watcon/proc42/contents html.)

Liebold, A.M., Saltus, R.W., Grauch, V.J.S., Lindsey, D.A., and Almquist, C.L., 1987, Mineral resources of the Sabinoso Wilderness Study Area, San Miguel County, New Mexico: U.S. Geological Survey Bulletin 1733-A, 26 p. (Also available at http://pubs.usgs.gov/bul/1733a/report.pdf.)

Lisenbee, Alvis, 2003, Preliminary geologic map of the Las Vegas NW 7.5-minute quadrangle, San Miguel County, New Mexico: New Mexico Bureau of Geology and Mineral Resources, Open-File Map OFGM-78. (Also available at http://geoinfo.nmt.edu/publications/maps/geologic/ofgm/details.cfml?Volume=78.)

Lohman, S.W., 1972, Ground-Water Hydraulics: U.S. Geological Survey Professional paper 708, 70 p. (Also available at http://pubs.er.usgs.gov/publication/pp708.)

Longworth, J.W., Valdez, J.M., Magnusen, M.L., Albury, E.S., and Keller, Jerry, 2008, New Mexico water use by categories 2005: New Mexico Office of the State Engineer, Technical Report 52, 111 p., accessed September 12, 2011, at http://www.ose.state nm.us/PDF/Publications/Library/TechnicalReports/TechReport-052.pdf.

Lucas, S.G., and Hayden, S.N., 1991, Type section of the Permian Bernal Formation and the Permian-Triassic boundary in north-central New Mexico: New Mexico Geology, v. 13, no. 1, p. 9–15. (Also available at http://geoinfo.nmt.edu/publications/periodicals/nmg/downloads/13/n1/nmg_v13_n1_p9.pdf.)

Lucas, S.G., and Hunt, A.P., 1987, Stratigraphy of the Anton Chico and Santa Rosa Formations, Triassic of east-central New Mexico: Arizona-Nevada Academy of Science Journal, v. 22, no. 1, p. 21–33.

Lucas, S.G., and Hunt, A.P., 1989, Revised Triassic stratigraphy in the Tucumcari basin, east-central New Mexico in Lucas, S.G., and Hunt, A.P., eds., Dawn of the age of dinosaurs in the American southwest: Albuquerque, N. Mex., New Mexico Museum of Natural History, p. 150–170. (Also available at http://nmnaturalhistory.org/assets/files/Bulletins/DawnAgeDinos/dawn_9_lucas.pdf.)

Maker, H.J., Derr, P.S., Anderson, J.U., and Link, V.G., 1972, Soil associations and land classification for irrigation, San Miguel County: New Mexico State University Agricultural Experiment Station Research Report 221, 44 p.

Martinez, F.J., 1990, Mora-San Miguel regional water study and forty year plan: Prepared for the Mora-San Miguel Water Plan Committee, on file at the Library of the New Mexico Office of the State Engineer, 193 p.

Matherne, A.M., and Stewart, A.M., 2012, Groundwater-well data of San Miguel County, New Mexico, 1970–2010: U.S. Geological Survey Data Series 686, 3 p., 2 tables. (Also available at http://pubs.er.usgs.gov/publication/ds686.)

Mattingly, B.E., 1990, A hydrogeologic evaluation of the Upper Pecos ground water basin in the vicinity of the Glorieta Baptist Conference Center, Glorieta, New Mexico: New Mexico Office of the State Engineer Technical Division Hydrology Report 90-1, 12 p., accessed September 12, 2011, at http://www.ose.state nm.us/PDF/Publications/Library/HydrologyReports/TDH-90-1.pdf.

McAda, D.P., and Wasiolek, Maryann, 1988, Simulation of the regional geohydrology of the Tesuque aquifer system near Santa Fe, New Mexico: US Geological Survey Water-Resource Investigations Report 87–4056. (Also available at http://pubs.usgs.gov/wri/1987/4056/report.pdf.)

Mercer, J.W., and Lappala, E.G., 1972, Ground-water resources of the Mora River drainage basin, western Mora County, New Mexico: New Mexico Office of the State Engineer Technical Report 37, 91 p. (Also available at http://www.ose.state nm.us/PDF/Publications/Library/TechnicalReports/TechReport-037.pdf.)

MJDarrconsult, Inc., 2003, Geohydrologic investigation report—Proposed "Santa Fe Mountain Ranches" subdivision, San Miguel County, New Mexico: Prepared for Santa Fe Mountain Ranches, on file at the Library of the New Mexico Office of the State Engineer, 17 p.

Molzen–Corbin & Associates, 1980, Wastewater re-use study: Prepared for the City of Las Vegas, N. Mex., on file at the Library of the New Mexico Office of the State Engineer, 10 p.

Molzen–Corbin & Associates and Lee Wilson and Associates, 1985, Las Vegas, New Mexico, water supply master plan status report: On file at the Library of the New Mexico Office of the State Engineer, 13 p., 13 apps.

Murray, C.R., 1943, Memorandum on ground-water conditions near the CAA Airfield northeast of Las Vegas, New Mexico: Prepared for the New Mexico Office of the State Engineer, U.S. Geological Survey technical memorandum, on file at the Library of the New Mexico Office of the State Engineer, 2 p.

Murray, C.R., 1944, Memorandum on geology of Peterson Reservoir, Las Vegas, New Mexico: Prepared for the New Mexico Office of the State Engineer, U.S. Geological Survey technical memorandum, on file at the Library of the New Mexico Office of the State Engineer, 3 p.

National Eutrophication Survey, 1977, Report on Conchas Reservoir, San Miguel County, New Mexico: U.S. Environmental Protection Agency, Working Paper No. 819, 46 p.

National Weather Service, 2011, U.S. Drought Monitor, New Mexico Drought Summary, September 27, 2011: accessed October 3, 2011, at http://www.srh noaa.gov/abq/?n=drought.

Natural Resources Conservation Service, 2011a, New Mexico SNOTEL sites: accessed June 25, 2011, at http://www.wcc.nrcs.usda.gov/snotel/New_Mexico/new_mexico html.

Natural Resources Conservation Service, 2011b, What is snow water equivalent?: accessed September 25, 2011, at http://www.or nrcs.usda.gov/Snow/about/swe html.

Natural Resources Conservation Service, [n.d.], Rapid watershed assessment—Western Estancia watershed: U.S. Department of Agriculture, 38 p., accessed September 2, 2011, at http://www.nm.nrcs.usda.gov/soils/watershed/RWAs/Western_Estancia.pdf.

New Mexico Environment Department, 2009, Gallinas River watershed monitoring, preliminary report: Santa Fe, N. Mex., New Mexico Environment Department Surface Water Quality Bureau, 20 p., accessed September 25, 2011, at ftp://ftp.nmenv.state.nm.us/www/swqb/MAS/Surveys/GallinasWatershed-2007.pdf.

New Mexico Environment Department, 2010, Gallinas River watershed monitoring, final report: Santa Fe, N. Mex., New Mexico Environment Department Surface Water Quality Bureau, 26 p., accessed September 25, 2011, at ftp://ftp nmenv.state nm.us/www/swqb/MAS/Surveys/GallinasMonitoringReport2010.pdf.

New Mexico Environment Department, 2011, Total maximum daily loads: accessed September 25, 2011, at http://www.nmenv.state.nm.us/swqb/TMDL.

New Mexico Office of the State Engineer, 2005, The state of the river—The status of the adjudication of the Pecos River: Santa Fe, N. Mex., New Mexico Office of the State Engineer Interstate Stream Commission, 37 p., accessed June 25, 2011, at http://www.ose.state.nm.us/water-info/legal/Pecos-RioGallinas/pecosRiverAdjudication.pdf.

New Mexico Office of the State Engineer, 2011a, Acequias: accessed September 25, 2011, at http://www.ose.state.nm.us/isc_acequias html.

New Mexico Office of the State Engineer, 2011b, Basins and programs: accessed September 25, 2011, at http://www.ose.state.nm.us/isc_basins_programs html.

New Mexico Office of the State Engineer, 2011c, Water rights research system—Areas abstracted into the database: New Mexico Water Rights Reporting System, New Mexico Office of the State Engineer disclaimer, accessed on October 3, 2011, at http://www.ose.state nm.us/PDF/Maps/WATERS-Abstract.pdf.

New Mexico Office of the State Engineer, 2011d, Underground water basins in New Mexico: accessed September 25, 2011, at http://www.ose.state nm.us/PDF/Maps/underground_water.pdf.

New Mexico Soil Conservation Service, 1994, Gallinas River Watershed natural resource plan: Santa Fe, N. Mex., New Mexico State Engineer Office Miscellaneous Documents No. 1297, on file at the Library of the New Mexico Office of the State Engineer, v. 1, 37 p., 2 apps.

New Mexico State Engineer Office, 1960, Progress report repair and improvement of Guadalupe and San Miguel County dams and canals: Santa Fe, N. Mex., New Mexico State Engineer Office Miscellaneous Documents No. 792, on file at the Library of the New Mexico Office of the State Engineer, 27 p.

New Mexico State Engineer Office, 1975, San Miguel County water resources assessment for planning purposes: Santa Fe, N. Mex., New Mexico State Engineer Office Interstate Stream Commission, 31 p.

New Mexico State Engineer Office, 1991, Pecos River Stream system hydrographic survey report, Gallinas River section, irrigation uses tracts GR–1A.a thru GR–17.104: Santa Fe, N. Mex., on file at the Library of the New Mexico Office of the State Engineer, v. 1, 487 p.

New Mexico State Engineer Office, 1997, Galisteo drainage basin water use inventory basic data report: Santa Fe, N. Mex., New Mexico State Engineer Office Miscellaneous Documents No. 489, on file at the Library of the New Mexico Office of the State Engineer, v. 1, 155 p.

Northrup, S.A., Sullwold, H.H., MacAlpin, A.J., and Rogers, C.P., 1946, Geologic maps of a part of the Las Vegas Basin and of the foothills of the Sangre de Cristo Mountains, San Miguel and Mora Counties, New Mexico: U.S. Geological Survey, Oil and Gas Investigations Map OM-54, scale 1:253,440. (Also available at http://ngmdb.usgs.gov/ngm-bin/ILView.pl?sid=5414_1.sid&vtype=a.)

Oregon State University, 2011, Analysis techniques—Flow duration analysis: Streamflow Evaluations for Watershed Restoration Planning and Design, accessed October 15, 2011, at http://streamflow.engr.oregonstate.edu/analysis/flow/index.htm.

Parker, J.T., and Pool, D.R., 1998, Use of microgravity to assess the effects of El Niño on ground-water storage in southern Arizona: U.S. Geological Survey Factsheet FS–060–98, 1 p. (Also available at http://pubs.usgs.gov/fs/FS-060-98/.)

Pool, D.R., and Schmidt, Werner, 1997, Measurement of ground-water storage change and specific yield using the temporal-gravity method near Rillito Creek, Tucson, Arizona: U.S. Geological Survey Water-Resources Investigations Report 97–4125, 30 p. (Also available at http://az.water.usgs.gov/pubs/WRIR97-4125intro html.)

Read, C.B., Wilpolt, R.H., Andrews, D.A., Summerson, C.H., and Wood, G.H., 1944, Geologic map and stratigraphic sections of Permian and Pennsylvanian rocks of parts of San Miguel, Santa Fe, Sandoval, Bernalillo, Torrance, and Valencia Counties, north-central New Mexico: U.S. Geological Survey Oil and Gas Investigations Map OM–21, scale 1:190,080. (Also available at http://ngmdb.usgs.gov/Prodesc/proddesc_5364 htm.)

Reynolds, C.B., and Reynolds, I.B., 2004, Petroleum possibilities of the Cuervo trough, Tucumcari Basin, New Mexico: New Mexico Geology, v. 26, no. 3, p. 83–89. (Also available at http://geoinfo nmt.edu/publications/periodicals/nmg/downloads/26/n3/nmg_v26_n3_p83.pdf.)

Robinson, W.P., 1995, Innovative administrative, technical, and public involvement approaches to environmental restoration at an inactive lead-zinc mining and milling complex near Pecos, New Mexico, *in* Waste Management Conference, Tucson, Ariz., 1995, Proceedings: University of Arizona/DOE/WEC, accessed September 14, 2011, at http://www.sric.org/mining/docs/Pecos html.

Rushton, Ken, 2007, Representation in regional models of saturated river-aquifer interaction for gaining/losing rivers: Journal of Hydrology, v. 334, p. 262–281. (Also available at http://www.sciencedirect.com/science/article/pii/S0022169406005440.)

Saavedra, Paul, 1987, Surface water irrigation organizations in New Mexico: New Mexico State Engineer Office, Report TDDC–87–2, accessed August 31, 2011, at http://www.nmacequiacommission.state nm.us/Publications/ose-acequia-rpt1987.pdf.

Schlaikjar, E.M., 1971, Geologic and ground-water report Variadero–Conchas Lake Area: Prepared for Ranchos Lake Conchas, Inc., on file at the Library of the New Mexico Office of the State Engineer, 16 p.

Searcy, J.K., 1959, Flow-duration curves—Manual of Hydrology, Part 2. Low-flow techniques: U.S. Geological Survey Water-Supply Paper 1542-A, 33 p. (Also available at http://pubs.usgs.gov/wsp/1542a/report.pdf.)

Shomaker, J.W., 1975a, Ground water conditions, Pendaries area: Prepared for Drissel and Associates, on file at the Library of the New Mexico Office of the State Engineer, 28 p.

Shomaker, J.W., 1975b, Ground water conditions, Hermit's Peak Ranch area: Prepared for Drissel and Associates, on file at the Library of the New Mexico Office of the State Engineer, 13 p.

Shomaker, J.W., 1976, Availability of ground water in Sec 8 & 9, T13N, R26E: Prepared for E.M. Wilson Enterprises, on file at the Library of the New Mexico Office of the State Engineer, 17 p.

Skotnicki, S.J., 2003, Preliminary geologic map of the Las Vegas quadrangle, San Miguel County, New Mexico: New Mexico Bureau of Geology and Mineral Resources, Open-File Map OF-GM-72, 5 p. (Also available at http://geoinfo.nmt.edu/publications/maps/geologic/ofgm/downloads/72/Las_Vegas_Report.pdf.)

Soil Conservation Service, 1959, Work plan for watershed protection and flood prevention, Pecos Arroyo Watershed: U. S. Department of Agriculture, 30 p.

Sorensen, Earl, and Gonzales, Ed, 1959, Reconnaissance of irrigated areas along the Pecos River between San Jose and Puerto de Luna, January 1959: Santa Fe, N. Mex., New Mexico Office of the State Engineer Miscellaneous Documents No. 1155, 30 p. (Also available at http://www.ose.state.nm.us/PDF/Publications/Library/HistoricalReports/1959-SEO-PecosRiverBetweenSanJose-PuertodeLuna.pdf.)

Souder, Miller & Associates, 1999, Geohydrologic report—Santa Fe Trail Ranch, San Miguel County, New Mexico: Prepared for Santa Fe Trail Ranch II, Inc., on file at the Library of the New Mexico Office of the State Engineer, 22 p., 4 apps.

Souder, Miller & Associates, 2010, Draft preliminary engineering report, water supply project, El Creston Mutual Domestic Water Consumers Association, San Miguel County, New Mexico: Prepared for El Creston Mutual Domestic Water Users Association, 21 p., accessed October 12, 2011, at http://elcreston.org/Welcome_files/El_Creston%20Draft%20PER%20Alternatives%20Considered.pdf.

Spiegel, S.J., 1956a, Geology and ground water possibilities in the vicinity of Las Vegas, New Mexico: Prepared for Public Service Company of New Mexico, on file at the Library of the New Mexico Office of the State Engineer, 9 p.

Spiegel, Zane, 1956b, Availability of ground water for North San Ysidro Domestic Water Consumers' Association: Santa Fe, N. Mex., New Mexico Office of the State Engineer, on file at the Library of the New Mexico Office of the State Engineer, 1 p.

Stonestrom, D.A., Prudic, D.E., Walvoord, M.A., Abraham, J.D., Stewart-Deaker, A.E., Glancy, P.A., Constantz, Jim, Laczniak, R.J., and Andraski, B.J., 2007, Focused ground-water recharge in the Amargosa Desert Basin, *in* Stonestrom, D.A., Constantz, Jim, Ferré, Ty, and Leake, Stanley, eds., Ground-water recharge in the arid and semiarid southwestern United States: U.S. Geological Survey Professional Paper 1703–E, p. 107–136. (Also available at http://pubs.usgs.gov/pp/pp1703/e/pp1703e.pdf.)

Summers, W.K., 1976, Catalog of thermal waters in New Mexico: New Mexico Bureau of Mines and Mineral Resources, Hydrologic Report 4, 79 p. (Also available at http://www.osti.gov/geothermal/purl.cover.jsp?purl=/7327302-Ocz4ql/native/.)

Surface Water Quality Bureau, 2001, Lake water quality monitoring, trophic state evaluation, and standards assessments for McAllister Lake, Storrie Lake, Santa Rosa Reservoir and Blue Hole: Santa Fe, N. Mex., New Mexico Environment Department Water Quality Assessments for Selected New Mexico Lakes, 27 p., accessed August 15, 2011, at http://www.nmenv.state nm.us/swqb/Surveys/LakeWaterQualityAssessments2001.pdf.

Surface Water Quality Bureau, 2004a, Water quality survey summary for the Upper Pecos River Watershed Part I (Between headwaters and Villanueva State Park), 2001: Santa Fe, N. Mex., New Mexico Environment Department, 21 p., accessed August 16, 2011, at http://www nmenv.state. nm.us/swqb/Surveys/UpperPecosPartI-2001.pdf.

Surface Water Quality Bureau, 2004b, Water quality survey summary for the Upper Pecos River Watershed, Part III (Between Tecolote Creek and Sumner Reservoir), 2001: Santa Fe, N. Mex., New Mexico Environment Department, 15 p., accessed August 16, 2011, at http://www nmenv.state. nm.us/swqb/Surveys/UpperPecosPartIII-2001.pdf.

Surface Water Quality Bureau, 2006, Lake water quality monitoring, trophic state evaluation, and standards assessments for Conchas Reservoir, Ute Reservoir, Upper and Lower Charette Lakes, Springer Lake, Lake Maloya, Lake Alice, Laguna Madre, Stubblefield Lake, Maxwell 12, 13, and 14, Lower and Upper Shuree Ponds in New Mexico: Santa Fe, N. Mex., New Mexico Environment Department Water Quality Assessments for Selected New Mexico Lakes, 58 p., accessed August 18, 2011, at ftp://ftp nmenv. state.nm.us/www/swqb/MAS/Surveys/Lakes-2006.pdf.

Surface Water Quality Bureau, 2008, Water quality survey summary for the Canadian River Tributaries (Vermejo River, Ocate Creek, and Mora River), 2002: Santa Fe, N. Mex., New Mexico Environment Department, 27 p., accessed August 16, 2011, at ftp://ftp nmenv. state.nm.us/www/swqb/MAS/Surveys/CanadianTribs-2002SurveyReport.pdf.

Surface Water Quality Bureau, 2009, Gallinas Watershed thinning monitoring preliminary report: Santa Fe, N. Mex., New Mexico Environment Department, 20 p., accessed August 16, 2011, at ftp://ftp nmenv.state.nm.us/www/swqb/MAS/Surveys/GallinasWatershed-2007.pdf.

Surface Water Quality Bureau, 2010a, Water quality survey summary for the Canadian River and select tributaries (Canadian headwaters to the Texas border and Cimarron River, Conchas River, Ute Creek, and Revuelto Creek), 2006: Santa Fe, N. Mex., New Mexico Environment Department, 33 p., accessed August 16, 2011, at ftp://ftp nmenv.state nm.us/www/swqb/MAS/Surveys/CanadianRiver-2006SurveyReport.pdf.

Surface Water Quality Bureau, 2010b, Gallinas Watershed thinning monitoring final report: Santa Fe, N. Mex., New Mexico Environment Department, 26 p., accessed August 16, 2011, at ftp://ftp.nmenv.state.nm.us/www/swqb/MAS/Surveys/GallinasMonitoringReport2010.pdf.

The People of San Miguel County and Communitas-Tierra y Gente, 2004, San Miguel County—Comprehensive plan, 2004–2014: 109 p., accessed June 13, 2011, at http://www.smcounty net/Public%20Notices/Draft%20Plan%20October%202003%20.pdf.

Thomson, Bruce, and Ali, Abdul-Mehdi, 2008, Water resources assessment of the Sapello River: Albuquerque, N. Mex., University of New Mexico, Water Resources Program, 45 p., accessed May 7, 2011, at http://repository. unm.edu/bitstream/handle/1928/6743/WR573_2008_Report.pdf?sequence=1.

Tolisano, Jim, Sharman, Jim, Brytowski, Jamie, and Friend, Harold, 1993, Environmental conditions, terrain management and geohydrology report for the Santa Fe Trail Ranch, San Miguel County, New Mexico: Prepared for Select Trading Group, Inc., on file at the Library of the New Mexico Office of the State Engineer, 32 p.

Trauger, F.D., 1972, Ground water in east-central New Mexico, *in* Kelley, V.C., and Trauger, F.D., eds., East-central New Mexico—New Mexico Geological Society, 23rd field conference, September 28–30, 1972: New Mexico Geological Society, p. 201–207. (Also available at http://nmgs.nmt.edu/publications/guidebooks/downloads/23/23_p0201_p0207.pdf.)

University of New Mexico, 2012, New Mexico's major reservoirs—An overview: Water matters! Background on selected water issues for members of the 51st New Mexico State Legislature 1st Session, 2012, p. 16-1–16-18.

U.S. Army Corps of Engineers, 2009, Environmental assessment for the Acequia Del Llano Rehabilitation Project, San Miguel County, New Mexico, Section 1113 Water Resources Development Act: U.S. Army Corps of Engineers, 36 p., accessed August 15, 2011, at http://www. spa.usace.army.mil/Portals/16/docs/environmental/fonsi/Del_Llano_Acequia_FEA.pdf.

U.S. Bureau of Reclamation, 1979, Tucumcari Project-Conchas Dam modification—Concluding report: Amarillo, Tex., U.S. Department of the Interior, on file at the Library of the New Mexico Office of the State Engineer, 22 p.

U.S. Census Bureau, 2010, 2010 population finder: accessed September 25, 2011, at http://www.census.gov/popfinder/.

U.S. Department of Agriculture, 2007, The census of agriculture, county profiles, New Mexico: National Agricultural Statistic Service, accessed September 25, 2011, at http://www.agcensus.usda.gov/Publications/2007/Online_Highlights/County_Profiles/New_Mexico/index.asp.

U.S. Environmental Protection Agency, 2011a, New Mexico water quality assessment report (2010)—Assessed waters of New Mexico by watershed: accessed September 25, 2011, at http://iaspub.epa.gov/waters10/attains_state.control?p_state=NM.

U.S. Environmental Protection Agency, 2011b, Superfund sites—Terrero Mine site information: accessed September 25, 2011, at http://cfpub.epa.gov/supercpad/cursites/csitinfo.cfm?id=0604054.

U.S. Environmental Protection Agency, 2011c, Superfund sites—El Molino Mill site information: accessed September 25, 2011, at http://cfpub2.epa.gov/supercpad/cursites/csitinfo.cfm?id=0600944.

U.S. Environmental Protection Agency, 2011d, Superfund sites—East Pecos site information: accessed September 25, 2011, at http://cfpub.epa.gov/supercpad/cursites/csitinfo.cfm?id=0605422.

U.S. Geological Survey, 1978, Gallinas River seepage investigation, *in* Water resources data for New Mexico, water year 1977: U.S. Geological Survey Water Data Report NM-77-1, p. 520–521.

U.S. Geological Survey, 2006, Gallinas Creek seepage investigation, *in* Miller, L.K., and Styles, J.A., Water resources data, New Mexico, water year 2005: U.S. Geological Survey Water Data Report NM-05-1, p. 414. (Also available at http://pubs.usgs.gov/wdr/2005/wdr-nm-05-1/.)

U.S. Geological Survey, 2011a, Floods—Recurrence intervals and 100-year floods: USGS Water Science for Schools, accessed October 15, 2011, at http://ga.water.usgs.gov/edu/100yearflood.html.

U.S. Geological Survey, 2011b, U.S. Geological Survey national geologic names lexicon (GEOLEX): accessed October 3, 2011, at http://ngmdb.usgs.gov/Geolex/geolex_home html.

Western Regional Climate Center, 2011, Western U.S. historical summaries—New Mexico: accessed March 24, 2011, at http://www.wrcc.dri.edu/summary/Climsmnm html.

White, W.E., and Kues, G.E., 1992, Inventory of springs in the State of New Mexico: U.S. Geological Survey Open-File Report 92–118, 251 p. (Also available at http://pubs.usgs.gov/of/1992/0118/report.pdf.)

Wilson, B.C., 2003, Water use by categories in New Mexico counties and river basins, and irrigated acreage in 2000: New Mexico Office of the State Engineer, Technical Report 51, 164 p. (Also available at http://www.ose.state.nm.us/PDF/Publications/Library/TechnicalReports/TechReport-051.pdf.)

Woodward, L.A., and Snyder, D.O., 1976, Structural framework of the southern Raton Basin, New Mexico, *in* Ewing, R.C., and Kues, B.S., eds., Vermejo Park, northeastern New Mexico—New Mexico Geological Society, 27th annual field conference, September 30–October 1–2, 1976: New Mexico Geological Society, p. 125–127.